Deposition of Atmospheric Pollutants

Deposition
of
Atmospheric Pollutants

*Proceedings of a Colloquium held at Oberursel/Taunus,
West Germany, 9-11 November 1981*

edited by

H.-W. GEORGII

Institute for Meteorology and Geophysics, University of Frankfurt, Germany

and

J. PANKRATH

Umweltbundesamt, Berlin (West), Germany

Springer-Science+Business Media, B.V.

Library of Congress Cataloging in Publication Data

Main entry under title:

Deposition of atmospheric pollutants.

Includes index.
1. Acid rain—Congresses. 2. Precipitation scavenging—Congresses.
I. Georgii, H.-W. II. Pankrath, Jürgen.
TD196.A25D46   1982   628.5'3   82–7669
                                             AACR2
ISBN 978-94-009-7866-9   ISBN 978-94-009-7864-5 (eBook)
DOI 10.1007/978-94-009-7864-5

TABLE OF CONTENTS

PREFACE

The problem of "acid precipitation" has been recognized with growing concern in many industrialized countries. The incorporation of pollutants into cloud and rain elements and their transfer to the ground by "wet deposition" are dominant mechanisms leading to a self-cleansing of the troposphere but, on the other hand, to hazards to the soil, vegetation and forests. The influence of orographic and meteorological parameters and of the regional distribution of precipitation on the deposition of pollutants are insufficiently known factors.

During previous years, several projects and analyses have been initiated to improve our knowledge on the dry and wet deposition of pollutants and on the mechanisms of transport of gaseous and particulate components from the atmosphere to the ground. Research activities have been supported in different fields and it appeared not only useful but necessary to bring the different research-groups together to endorse the communication and cooperation between scientists in the related fields. A symposium was arranged in Oberursel/ Taunus in November 1981 to discuss the results of experimental and theoretical work in the field of deposition and to gain a better understanding of each other's methods, experience and observations.

The proceedings presented in this volume permit a fair survey of the present-day knowledge and will be a useful tool for all working in this field.

The meeting would not have been possible without the financial support of the German Environmental Agency (Umwelt-bundesamt).

The preparation of the symposium and of the publication of the contributions caused a considerable amount of work. I would like to express my gratitude to Miss C. Perseke, Mr. E. Rohbock and Mrs. H. Wallenwein for their continuous assistance.

Hans-Walter Georgii

Introduction

COMMENTS ON THE INVESTIGATION OF THE DEPOSITION OF ATMOSPHERIC POLLUTANTS

Dr. J. Pankrath

This colloquium is part of a R&D-programme which is funded by the Federal Environmental Agency in the framework of the Environmental Research Programme of the Ministry of the Interior. The research is being carried out under the leadership of Professor H.-W. Georgii of the Institute of Meteorology and Geophysics of the University of Frankfurt and is concerned with the deposition of atmospheric pollutants with special emphasis on wet deposition. From an environmental point of view the measurements of sulphates, nitrates and heavy metals in a precipitation chemistry network is of great importance, because these components pose a threat to ecology in regions well removed from industrialized areas.

We can generally speak of acid rain in this context being aware that there are other components such as heavy metals and organic compounds which are distributed over longer distances as well. Compared with other types of air pollution, acid rain is less well understood and of less immediate concern in the minds of the respective decision makers. Public awareness is now growing due to information about the menace of acid rain which, for instance, is said to be partly responsible for the damage of spruce forests also in Germany. The special problem of acid rain is its slow and insidious effect on the environment: changes are noticed only when they are irreversible.

In fact, it is misleading to believe that the use of high chimneys leads to air pollution being well mixed in the atmosphere, thus resulting in insignificant concentrations at ground levels. Moreover, while there have been major improvements, many sources of pollution are just as bad as ever. Acid rain is produced when sulphur dioxide and nitrogen oxides combine with oxygen in the air and with water vapour to form sulphuric acid and nitric acid. The SO_2 is

H.-W. Georgii and J. Pankrath (eds.), Deposition of Atmospheric Pollutants, 3–6.
Copyright © 1982 by D. Reidel Publishing Company.

derived mainly from the burning of fossil fuels that contain sulphur. Nitrogen oxides are a product of the chemical combination of oxygen and nitrogen at high furnace temperatures and in internal combustion engines. These gaseous emissions may be carried hundreds and even thousands of kilometres before the atmosphere is depleted by dry and wet deposition. The farther these emissions travel, the longer they are exposed to reactions of different pathways ending up with partly acid sulphates and nitrates. The deposition of these substances may change the chemistry of soil, making it less alkaline and more acidic; they may damage foliage that absorbs them; and they may change the chemical balance of vulnerable lakes, so that these can no longer support aquatic life.

The acidity of precipitation varies greatly from day to day but in the annual mean it is now at about a pH-value of 4.2 to 4.5 at rural sites in Germany. That means the precipitation is about 25 fold more acid than it would be if there were no air pollution but had a CO_2 content of about 340 ppm.

In the case of long range transport, the actual concentrations of air pollutants are generally small and nearly always below air quality levels. It is the long-term accumulation by deposition that is important.

The precise response of terrestrial ecosystems to acidic and other toxic substances is extremely difficult to assess and predict. For instance, serious changes have been noted in soil chemistry which are believed to be triggered by acid rain. These changes are reflected in the serious degradation of beech and spruce forests. Every environmental hazard has its own biological monitor. Forests act acid concentrators, that means acid reaching the forest ground annually is two (beech) to four (spruce) times the acid in the rain entering the forest canopy. This is so because foliage and bark capture and oxidize sulphur dioxide present in the atmosphere in very low concentrations. The resulting acid is then added to the acid aerosols already deposited in the canopy from the air. When it rains, both these acid sources are cleansed from the trees and reach the forest ground.

The increased acid input slows the breakdown of humus, and carbon and nitrogen stores in the soil are increasing. The increased nitrogen in the soil undergoes additional reactions that generate more acid, further stressing the trees, mainly by releasing inorganic aluminium into the soil solution. This aluminium is toxic to the fine roots in the soil's mineral layer, and they become damaged. The trees survive though they are severely stressed because the fine rootlets in the humus layer above are less effected by the organically complexed aluminium at that soil level. Although the trees' ultimate survival is in doubt, liming is recommended as a remedial measure to rectify damage done by the altered soil chemistries.

Despite numerous investigations, the acid rain problem is a controversial subject that encompasses scientific, economic, and political issues. The most obvious requirement is to achieve a sound scientific understanding of the problem. But this is hampered by the fact that available data and numerical calculations at the moment cannot obviate contrasting views.

There are disturbing, unresolved problems that make it difficult even to know what the true pH of rain, not influenced by anthropogenic activities, ought to be, although most scientists believe it to be at about 5.6. There are natural sources of acidifying substances emitted into the atmosphere as well as pollutants arising from combustion of fuels and other processes. Neutralizing substances present in the atmosphere also affect the acidity of rain. Because of the episodic nature of precipitation events it may require decades of information on the composition of rain before trends can be established unambiguously.

Rain contains elements important for plant growth, namely nitrogen and sulphur, as well as excessive acidity. Nitrogen and sulphur in rain are absorbed by foliage and by plant roots from the soil and are then used in support of plant productivity. Excessive acidity, however, interferes with physiological processes and can reduce growth and yield. There are, then, two opposing influences at work and the net effect seems to vary with species, stage of plant development, pattern of rainfall, soil nutrient supply and probably other factors as well. Because the biological processes are complex, it appears that long-term studies are needed to provide conclusive answers to the harmfulness of acid rain. In an analogous manner, the exact contribution of acid deposition versus natural processes in the acidification of lakes cannot be assessed conclusively.

One important question about acid deposition concerns the scale of effects that result from the deposition of substances from the atmosphere. Industrial societies now appear capable of inadvertently altering regional and even global atmospheric chemistry. Perhaps one of the most disturbing consequences of acid rain is that risks of uncertain magnitude may be imposed on the population of other nations. Being aware of this situation, the member countries of the Economic Commission for Europe (ECE) of the United Nations have signed in 1979 a convention in order to combat trans-

boundary air pollution fluxes. A scientific co-operative
programme, called EMEP, is part of this convention; the
programme is to proceed with the investigation into the
fundamentals of the mutual impact of other nation's
emissions. The USA and Canada are carrying out a comprehensive research programme in accordance with the memorandum of
intent on transboundary air pollution concluded in 1980.
There are some efforts in this direction in Germany as well,
in order to obtain sufficient information on which to base
effective action to prevent ecology suffering from harmful
impacts. But the main difficulty is not resolved as yet: it
culminates in the question: how can uneffective decisions
be avoided when strategies have to be developed with an
imperfect knowledge of the problem. This is a fundamental
issue that **requires** thorough discussion and **analysis** if the
decision makers are set to national environmental policies
to deal with the acid rain problem. Sufficient information
does not exist on the extent of damage, on the causes, and
on the transport/transformation mechanism that contribute
to acid deposition. Owing to this complexity, a linear
relationship between emission rates and deposition rates at
specific sites cannot be expected at all.

Concluding my short glance on the acid deposition problem,
I would like to invite the **assembly** to draw their attention
to the question of public acceptance of long term environmental and ecological risks. The research into the deposition of acid and, more general, of harmful substances can be
an adequate incentive.

Dry Deposition

FIELD MEASUREMENTS OF THE DRY DEPOSITION OF SMALL PARTICLES TO GRASS

J A Garland
AERE, Harwell

ABSTRACT

Deposition rates determine the extent and intensity of the effects of
atmospheric pollution. Previous measurements of the rate of dry deposition of small particles have suggested that there may be a systematic
difference between wind tunnel and field results. Three recent field
experiments on the dry deposition of lead, Aitken nuclei and 1.8 μm
monodisperse particles are discussed. They give deposition velocities
within a factor of three of the wind tunnel data. The larger deposition
velocities suggested in some other experiments are not consistent with
estimates of the atmospheric lifetime of small particles.

1. INTRODUCTION

Dry deposition is an important mechanism for the removal of certain
reactive gases and large particles from the atmosphere. The deposition
of smaller particles in the region 0.05 to 1 μm is of interest since
this region contains a large fraction by mass of the atmospheric aerosol.
In particular, the sulphate aerosol, some heavy metals, and many other
products of combustion are found in this size range, and the rate of
deposition influences the geographical distribution and intensity of the
effects of industrial emissions on plants, soil chemistry and crop
contamination.

Several investigators have studied the deposition of small particles to
smooth and rough surfaces and elements of foliage in wind tunnels, (eg
see Chamberlain, 1966; Clough, 1975; Möller and Schumann, 1970, Belot
et al, 1976). Results are usually expressed as the deposition velocity.

$$v_g = \frac{F}{\chi(Z_r)}$$

where F is the flux density of particles to the surface, and $\chi(Z_r)$ the
concentration measured at a suitable reference height, Z_r, within the
boundary layer above the surface. Wind tunnel experiments give direct
measurements of v_g to grass and similar surfaces, and permit deposition

9

H.-W. Georgii and J. Pankrath (eds.), Deposition of Atmospheric Pollutants, 9–16.
Copyright © 1982 by UKAEA

to forest to be predicted (Belot et al, 1976). For particles in the range 0.05 to 1 μm diameter the wind tunnel results indicate a minimum of v_g (eg see Fig 1) with values in the range 3×10^{-3} to 5×10^{-2} cm s^{-1} for a wide range of wind speeds and for surfaces as different as short grass and forest. Larger and smaller particles deposit more rapidly, due to increasing departure of particle trajectories from the gas flow occasioned by Brownian motion (for smaller) and inertia (for larger particles).

Field observations of the deposition velocity of particles in the 0.05 to 1 μm size range are difficult to perform and few are reported in the literature. They give more variable results and often indicate deposition velocities larger than 0.1 cm s^{-1}. Some published experiments are attempts to use micrometeorological methods. Everett et al (1979) measured the variation of the concentration of sulphate with height and deduced deposition velocities of 1 or 2 cm s^{-1}. Wesely et al (1977) found deposition velocities of 0.1 to 1 cm s^{-1} using an electrical charging device to detect particles of about 0.05 to 0.1 μm in an eddy correlation experiment.

Several trace elements commonly occur in the size region of interest, and direct measurements of their accumulation on grass or on artificial surfaces provide evidence of the deposition of small particles. Davidson and Friedlander (1978) considered the observed fluxes of lead to grass (which correspond to $v_g \sim 1$ cm s^{-1}). The lead,chiefly from motor exhaust,is predominantly sub-micron, but they explained the large v_g by the presence of small numbers of large particles. The same explanation may account for the deposition velocities to filter paper in the range 0.2 to 0.7 cm s^{-1} found by Cawse (1974) for several trace elements with submicron median diameters in the atmospheric aerosol. However the large values of deposition velocity for submicron particles observed in some field measurements have not been reconciled with the wind tunnel observations.

The large difference between some estimates of v_g obtained in field experiments and in the wind tunnel suggests that there may be an important difference between the mechanisms of deposition in the two circumstances. It has been suggested that such a difference might arise because of differences in the spectrum of turbulence in the two situations; certainly eddies of metre to kilometre scale exist in the atmosphere, while the eddie size in the wind tunnel is limited by the dimensions of the tunnel cross section and cannot usually exceed ~ 1 m. it is important to resolve this issue since the higher values of deposition velocity suggested in some field experiments would require an important change in our understanding of the lifetime and behaviour of the atmospheric aerosol.

Here we discuss the results of three recent experiments regarding the deposition of particles near the size of the minimum of the deposition velocity curve.

2.   DEPOSITION OF LEAD TO GRASS

Little and Wiffen (1978) measured the deposition of lead to grass beside the M4 motorway in West London, where the road carries about 90,000 vehicles per day. The freshly emitted lead in the exhaust smoke of vehicles on the motorway dominated the lead aerosol during the measurements. Trays of grass, grown in the laboratory, were placed on the ground on the verges and adjacent fields and exposed for periods of several hours. Simultaneous filter samples were taken to allow measurement of the lead concentration in air. After exposure samples of grass were cropped from each tray, wet ashed with nitric acid, and analysed for lead by atomic absorption spectrophotometry. Thus the deposition rate of lead to grass could be directly measured, and the deposition velocity could be deduced.

The deposition velocity decreased with distance from the edge of the motorway from 2 m to 18 m, but values at 18 m and 30 m were not significantly different. The high values close to the road were probably due to turbulence generated by the traffic. Evidence of the intensity of the turbulence was provided by cup anemometers placed at each measurement station. Those at 2 and 8 m from the roadway indicated windspeeds substantially higher than those at 18 and 30 m. Deposition velocities observed at 18 and 30 m are listed in Table 1, extracted from Table 4 of Little and Wiffen (1978).

Table 1 Deposition of lead to grass beside the M4 Motorway

Distance from edge of hard shoulder m	Deposition velocity, cm s^{-1} and Wind speed, m s^{-1}, in brackets					
	Run 1	Run 2	Run 3	Run 4	Run 5	Run 6
2	1.58 (3.5)	1.8 (3.75)	4.36 (5.0)	0.88 (1.65)	0.58 (3.0)	0.54 (2.5)
8			2.0 (2.5)	0.42 (0.9)	0.52 (1.5)	0.5 (0.8)
18				0.29 (0.85)	0.29 (1.3)	0.33 (0.75)
33				0.34 (0.85)	0.25 (1.4)	0.27 (0.8)

Mean of observations at 18 and 33 m: v_g = 0.30 cm s^{-1}

Wind speed = 1.0   m s^{-1}

The lead aerosol at this location has been studied by diffusion battery and electron microscopy (Little and Wiffen 1978), and by sampling with Andersen impactors (Little and Wiffen 1977). The first two instruments were used to investigate the size distribution of the numerous particles, less than 0.1 μm diameter, which contain most of the particulate lead. These measurements indicated volumetric mean diameters of 0.024 to 0.08 μm. Most frequently the mean diameter was 0.03 to 0.05 μm. The Andersen sampler showed a mean of 67% on the final filter, the remainder of the lead being distributed between the impactor stages. Table 2 shows these results, and also the effect of correcting the Andersen results for the observed collection of a small fraction of the sub-tenth-micron aerosol on the impactor stages. (M J Heard, private communication).

Table 2   Mean size distribution and expected deposition
velocity for lead near the M4 motorway

Andersen Sampler stage number	Equivalent Particle diameter μm	% of total deposit	Distribution corrected for deposition of fine aerosol	Deposition velocity cm s^{-1}
0	13	1.43	0.6	1.9
1	10	1.94	1.1	1.4
2	7.5	3.3	2.4	0.8
3	5.0	4.09	3.1	0.3
4	3.2	4.67	3.7	0.1
5	1.6	4.44	3.5	0.03
6	0.75	8.7	7.5	0.02
7	0.4	4.46	3.5	0.02
Backing filter	0.04	66.98	74.6	0.09

Mean expected deposition velocity = 0.13 cm s^{-1}

This information on the size distribution makes it possible to compare the deposition velocity observed in the field with the values obtained in wind tunnel measurements. The last column of Table 2 shows the details of the calculation of the deposition velocity to be expected in the field measurements. The deposition velocities assigned to the particles collected on each stage of the Andersen sampler, were obtained from Fig 1. The deposition velocity for the lead observed in the field, is about twice the value predicted using the wind tunnel results.

3.   DEPOSITION OF CONDENSATION NUCLEI

Deposition at the ground is accompanied by a decrease of concentration near the surface. The resulting concentration gradient in the lowest few metres of the atmosphere can be related to the rate of deposition, and this method has been used to investigate the deposition of SO_2 (eg Garland, 1977) and other gases. Gradients of temperature and humidity are often used to deduce the heat transfer and transpiration rate from crops and other surfaces.

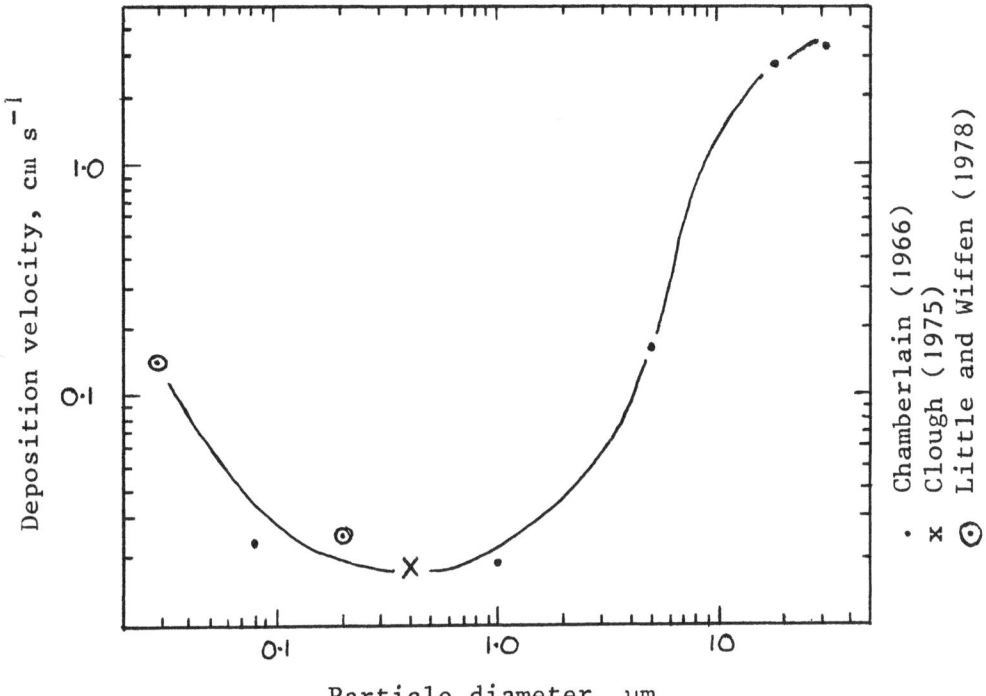

Fig 1   Deposition velocity of particles to grass measured in wind tunnel experiments at wind speeds of 2.5 to 3.0 m s^{-1}.

Garland and Cox (1982) used an automated Nolan-Polak condensation nucleus counter to measure the concentration of Aitken nuclei at heights of 0.2, 1 and 2 m above a grass field. These particles are present in numbers of order 10^3 to 10^5 per cm^3 in terrestrial air. The size distribution of the Aitken particles has been studied extensively, and shows certain features consistently. A large fraction of the total number is present in particles smaller than 0.03 μm diameter and at larger sizes the number of particles in each equal-logarithmic size interval declines rapidly with increasing particle size. To improve the definition of the size interval included in their measurements, Garland and Cox included a diffusion battery before the nucleus counter, so that particles smaller than ∿ 0.05 μm were removed from the air. In these conditions the counter responds chiefly to particles in the size interval of 0.05 to 0.2 μm diameter.

The concentration of condensation nuclei varies from time to time, reflecting changes in wind speed and direction, and the spatial distribution of combustion sources. Measurement periods were chosen to avoid occasions of rapid fluctuations in concentration, but the remaining variations made it necessary to average several days observations in order to detect the small difference in concentration between different measurement heights. In aggregate, the results showed a deficit of 2.7 ± 1% in

the concentration at 0.2 m in comparison with that at 1 m. No significant difference was detected between the 1 m and 2 m heights. The reader
is referred to Garland and Cox (1982) for an explanation of interpretation
of the results. They deduced that deposition velocity for 0.05 to 0.2 μm
particles to grass is 0.06 ± 0.03 cm s^{-1}. The mean wind speed was
$\sim$ 3 m s^{-1}. The expected deposition velocity according to Fig 1, was
in the region 0.02 to 0.05 cm s^{-1}.

4.   DEPOSITION OF 1.8 μm IRON OXIDE PARTICLES TO GRASS IN THE FIELD
      AND IN A WIND TUNNEL

Spherical particles of an oxide of iron (β-FeOOH) labelled with radioactive ^{59}Fe were made using a spinning top aerosol generator. Particles
prepared in this fashion have a density of about 2.5 g cm^{-3}. The
particles were 1.8 μm diameter, and their aerodynamic equivalent diameter
was therefore 2.8 μm. They were collected in an impinger, and the
resulting suspension was atomised above a grass field so as to produce
an aerosol of singly dispersed particles. The grass was about 10 cm tall.
The atomiser was located 50 cm above the surface, about 25 m upwind of an
array of air samplers. Subsequently, samples of the grass close to the
air samplers were collected, and their content of ^{59}Fe was assessed by
γ-scintillation counting.

Similar experiments were carried out in a portable wind tunnel, described
by Garland (1979). This could be located in the same field so that the
grass surface of the field formed the floor of the working section.
Batches of the same particles were atomised in the wind tunnel and filter
samples of the air, and grass samples, were collected 15 m down wind.

The results, shown in Table 3, show a considerable spread of values. A
small part of the variation may be due to differences in wind speed.
There is no indication in the results that the deposition velocity is
higher in the field than in wind tunnel measurements. The expected
deposition velocity for the particles used is 0.07 cm s^{-1} (Fig 1).

Table 3.   Deposition velocity for 1.8 μm iron oxide
                    particles to grass

Date, 1981	Method	Wind speed m s^{-1}	Deposition velocity cm s^{-1}
21 May	Field	2.5	0.05
1 June	Wind tunnel	3.5	0.18
17 August	Wind tunnel	3.5	0.12
23 September	Field	3	0.12

5. DISCUSSION

The findings of the three investigations quoted above are summarised in Table 4. The deposition velocities for lead and Aitken nuclei, measured in the field experiments, are about a factor of two larger than the values expected from wind tunnel measurements. The wind tunnel measurements were made at 250 to 300 cm s^{-1} at a height of 15 cm, equivalent to $\sim$ 400 cm s^{-1} at 1 m in the field. The field measurements were conducted at average wind speeds of $\sim$ 1 m s^{-1} for lead, and 3 m s^{-1} for Aitken nuclei. The deposition velocity is expected to increase with wind speeds, but by less than the ratio in wind speed. On the other hand, the grass in the field experiments was probably two or three times taller than that in the wind tunnel, and this difference might account for a small increase in v_g in the field. In conclusion, it is not possible to account for the significant difference between these field measurements for lead and Aitken nuclei and wind-tunnel experiments, but the difference is not very large.

Table 4. Summary of field observations of deposition velocity

| Aerosol | Deposition velocity, cm s^{-1} | |
	Observed	Predicted from wind tunnel studies
Exhaust lead aerosol, mainly $\sim$ 0.04 μm	0.30	0.13
Aitken nuclei $\sim$ 0.05–0.2 μm	0.06 ± 0.03	0.02 – 0.05
1.8 μm iron oxide, density = 2.5 g cm^{-3}	0.05 – 0.12	0.07

The third experiment allows a direct comparison between field and wind tunnel for similar wind speeds and surface conditions. The results are scattered, and more data is required, but the present results suggest agreement within a factor of two.

On the whole, the results support the application of wind tunnel results in field conditions, with an uncertainty of a factor of $\sim$ 3.

Dry deposition, precipitation and atmospheric mixing determine the atmospheric residence time of particulate matter. Material released at or near the ground enters the boundary layer, about 1 km deep on average. Consideration of the behaviour of various radioactive tracers (Martell and Moore 1974) and of the man-made sulphate aerosol (Rodhe 1978) leads to estimates of $\sim$ 3 days to 1 week for mean residence times of substances which reside chiefly on submicron particles. If removal by precipitation is neglected, an upper limit of $\sim$ 0.3 cm s^{-1} is obtained for the mean deposition velocity for such particles. As precipitation removes substantial quantities of these substances, the mean deposition velocity is probably a few times smaller. It is difficult to compare such deductions with short term experimental results because of the affects of size distribution and variable atmospheric mixing, but it is clear that there are important components of the atmospheric aerosol which have a mean deposition velocity no greater than $\sim$ 0.1 cm s^{-1}.

CONCLUSIONS

Three field experiments for particles in the size range 0.05 μm to 1.8 μm indicate deposition velocities of order 0.1 cm s^{-1} or less. These results are no more than a factor of 3 higher than wind tunnel measurements suggest. The larger deposition velocities suggested in some other experiments are not consistent with estimates of the atmospheric lifetime of small particles.

REFERENCES

Belot, Y., Baille, A., and Delmas, J.-L: 1976, Atmospheric Environment 10, pp59-98.

Cawse, P.A.: 1974, AERE-R 7669, HMSO, London.

Chamberlain, A.C.: 1966, Proc. Roy. Soc. A 296, pp45-70.

Clough, W.S.: 1975, Atmospheric Environment 9, pp111-1119.

Davidson, C.I., and Friedlander, S.K.: 1978, J. Geophys. Res. 83, pp2343-2352.

Everett, R.G., Hicks, B.B., Berg, W.W., and Winchester, J.W.: 1979, Atmospheric Environment 13, pp931-934.

Garland, J.A.: 1977, Proc. Roy. Soc. A 354, pp245-268.

Garland, J.A.: 1979, AERE-R 9452, HMSO, London.

Garland, J.A., and Cox, L.C.: 1982, Atmospheric Environment, to be published.

Little, P., and Wiffen, R.D.: 1977, Atmospheric Environment 11, pp437-447.

Little, P., and Wiffen, R.D.: 1978, Atmospheric Environment 12, pp1331-1341.

Martell, E., and Moore, N.E.: 1974, J, Rech. Atmos. 8, pp903-910.

Rodhe, H.: 1978, Atmospheric Environment 12, pp671-680.

Wesely, M.L., and Hicks, B.B.: 1977, Atmospheric Environment 11, pp561-563.

ON THE VERTICAL FLUX OF GASEOUS AMMONIA ABOVE WATER AND SOIL SURFACES

László Horváth
Institute for Atmospheric Physics, H-1675 Budapest P.O.B. 39.
Hungary

ABSTRACT. The vertical flux of ammonia gas has been determined above soil and water surfaces, in summer half-year, in Hungary. Above water surface the sign and the measure of the flux depends on the difference between the atmospheric ammonia level and the calculated equilibrium concentration. These concentration differences give a good correlation with the values of the measured flux. Above uncultivated soil surface deposition process takes place in most cases. It suggests that the atmospheric ammonia level generally exceeds the equilibrium concentration, determined by the chemical composition of soil. The average deposition velocity is 0.48 cm.s^{-1}.

1. INTRODUCTION

For the study of the atmospheric cycle of ammonia one has to exactly know the interaction between the ammonia and the natural surfaces (soil, water).

The role and importance of atmospheric ammonia gas are well known. Among other things ammonia controls the heterogeneous oxidation of sulfur dioxide (Scott and Hobbs, 1967) and ammonia is the only basic gas in the atmosphere able to neutralize the acids formed by oxidation of several trace gases (e.g. nitrogen dioxide, sulfur dioxide). According to Mészáros and Vissy (1974) ammonium can be identified as the dominant cation in the tropospheric background aerosol.

The ammonium/ammonia ratio in aquatic and terrestrial ecosystems depends on the concentration of hidrogen ion (H^+) considering the hydrolysis of ammonia:

$$\frac{(NH_4^+)_w}{(NH_3)_w} = (H^+)\frac{K_h}{K_w} , \qquad (1)$$

where $(NH_4^+)_w$ and $(NH_3)_w$ are the concentrations in the aquatic or terrestrial ecosystems, K_h is the equilibrium constant of hydrolysis, while K_w

17

H.-W. Georgii and J. Pankrath (eds.), Deposition of Atmospheric Pollutants, 17–22.

is the ion product of water. In the state of equilibrium the distribution of ammonia between the two phases is determined by Henry's law:

$$H_1 = \frac{(NH_3)_w}{(NH_3)} \; ,$$

(2)

where H_1 is constant and (NH_3) is the ammonia concentration in gas phase. For this reason from the point of view of ammonia two processes can take part at the surface of natural ecosystems namely the emission or the absorption. When the equilibrium atmospheric concentration $(NH_3)_{eq}$, determined by the physical and chemical state of water or soil, differs from the actual ammonia level (NH_3) in the air, absorption or emission takes place to approach the equilibrium. These processes result in ammonia flux to or from the surfaces. In this way the water and the soil may be a source or a sink of atmospheric ammonia gas.

The aim of this study has been to determine the deposition and emission processes for natural surfaces and correlate the measured vertical flux with the physical and chemical parameters of water or soil and air.

2. EXPERIMENTAL

For ammonia sampling the method of Ferm (1979) was used during which the effect of the ammonium particles was eliminated.

For measuring the ammonia flux above water and soil surfaces the gradient method was adapted. Deposition (or emission) was calculated by means of the following formula:

$$d = k_{2,1} \frac{c_2 - c_1}{h_2 - h_1} \; ,$$

(3)

where d is the deposition of ammonia, c_1 and c_2 are the ammonia concentrations at lower (h_1) and upper (h_2) level, respectively, while $k_{2,1}$ is the eddy diffusivity. One can see that in the case of $c_1 > c_2$ the sign of deposition is negative (i.e. emission occures). The height of the lower level was chosen to be 0.5 m while the higher one was 2.0 m in the case of water and 4.0 m for the soil surface.

Measurements above water surface were carried out at Lake Balaton (~ 600 km^2, average depth: 3 m) in Hungary using a small boat. The distance between the shore and the measuring point was at least 1 km. The average distance of the nearest shore downwind was 3.5 km.

In the case of soil surface the measurements were carried out in a suburb of Budapest in the garden of Institute for Atmospheric Physics above uncultivated, sandy soil, covered by short grass.

Both for water and soil the sampling period was 2-8 hours. The samples were taken in the day-time. The volume of air sampled was 0.5-1.5 m^3. After sampling the samples were analysed by means of an ammonia sensitive electrode (OP-NH$_3$-7113, RADELKIS) or a spectrophotometer (indophenol-blue method). Former determination was used at Lake Balaton while latter one above soil. The detection limit was found to be 0.05 $\mu g \cdot m^{-3}$ for a sampling volume of 1 m^3 with a relative standard deviation of 15 %.

In addition to ammonia measurements the ammonium plus ammonia content and the pH of the water, the temperature of air, water and soil (at 5 cm level), as well as the wind direction and the average wind speed (at both levels) were also detected. All the measurements were carried out in the summer half-year.

3. AMMONIA FLUX ABOVE WATER SURFACE

The equilibrium ammonia concentration above the water surface can be calculated from (1) and (2):

$$(NH_3)_{eq} = \frac{c_w}{\frac{(H^+)K_h H_1}{K_w} + H_1} \tag{4}$$

where c_w is the sum of the ammonium and ammonia concentrations in the water phase. However, the calculated equilibrium concentrations are much lower compared with the measured values (Lau and Charlson, 1977). The probably explanation of this discrepancy was given by Hales and Drewes (1979). According to their new solubility theory supported by direct laboratory measurements the ambient carbon dioxide gas reduces the solubility of ammonia to a large extent. The stationary state condition can be described by the following formula:

$$(NH_3)_{eq} = \frac{c_w |H_1 H_2 (CO_2) Q + 1|}{H_1 H_2 P(CO_2) + \frac{(H^+)K_h H_1}{K_w} + H_1} \tag{5}$$

where H_2 is the Henry's law constant for carbon dioxide, P and Q are constants at a given temperature while (CO_2) is the atmospheric concentration of carbon dioxide (considered to be 350 ppm in this paper). One can easily recognize that in the case of $(CO_2)=0$ the equation (5) yields (4).

Using the average parameters of Lake Balaton (pH=8.1, $c_w=0.029$ mg·l^{-1}) the concentrations calculated by means (5) are between 0.32 and 1.06 as a function of water temperature (see: Fig. 1). Because the background ammonia concentration in Hungary is 1.39 $\mu g \cdot m^{-3}$ in the summer half-year there is no essential difference between the atmospheric ammonia level and the calculated concentration (1.0) at the average water temperature (21 °C) during flux measurements. Hence, because of the temporal variation of the atmospheric concentration the equilibrium ammonia level may exceed the atmospheric level. Therefore the direction of the flux (deposition or emission) depends on the sign of the difference between the above mentioned concentrations. If the new solubility theory is suitable for this system the differences between the calculated and atmospheric concentra-

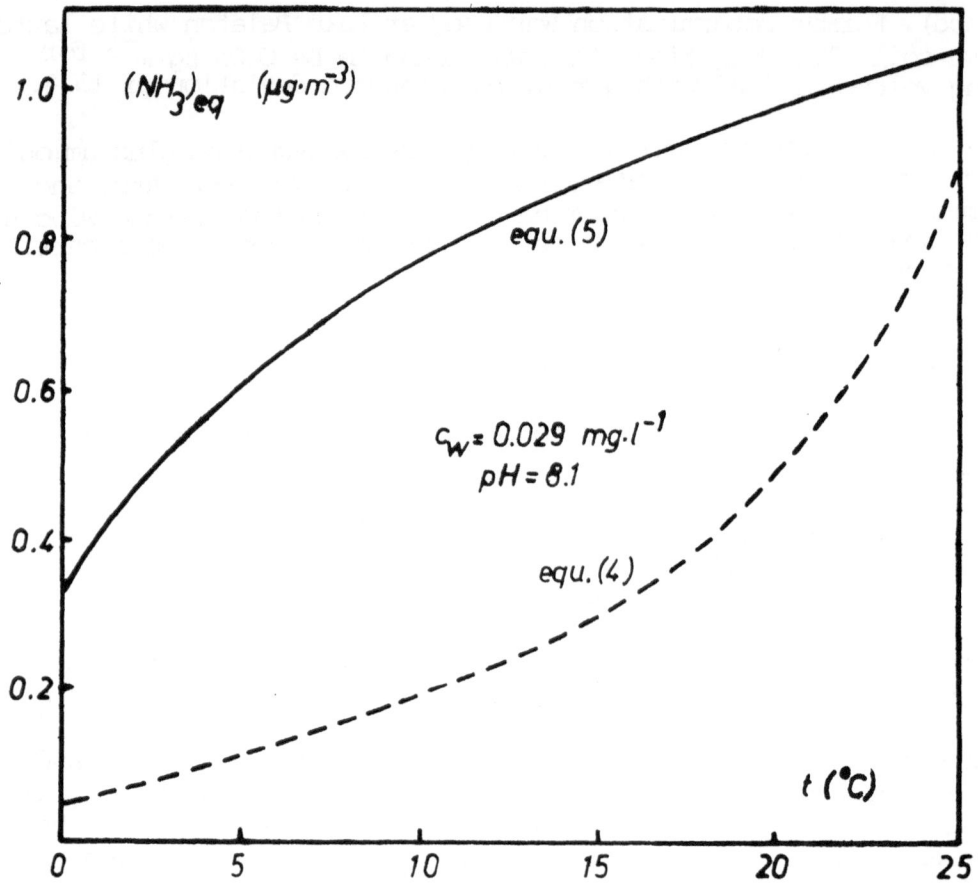

Fig. 1. Calculated equilibrium ammonia concentration above Lake Balaton
        as a function of water temperature.

tions (Δc) should give a correlation with the measured flux. On the basis
of 26 measurements the regression line is:

$$d = 0.0068 \, \Delta c - 0.0129, \tag{6}$$

where d is the rate of measured flux in $\mu g \cdot m^{-2} \cdot s^{-1}$, Δc is expressed in
$\mu g \cdot m^{-3}$ units. The correlation coefficient is $r=0.7$ at the probability
level of $p=0.1$ %. The average flux equals to $-0.002 \, \mu g \cdot m^{-2} \cdot s^{-1}$ with a maximum of 0.131 and a minimum of $-0.087 \, \mu g \cdot m^{-2} \cdot s^{-1}$. The average concentrations were 2.56 and 2.59 $\mu g \cdot m^{-3}$ at the lower and upper level, respectively. The average difference between upper and lower concentrations was
54 % independing of the sign of the flux. In 12 cases deposition occured
(positive flux) while in 14 cases emission was observed.

4. AMMONIA FLUX ABOVE SOIL SURFACE

During the measurement of ammonia flux the chemical composition of the
soil was not examined. As a first approximation it was supposed that the
pH and the ammonium concentration in the soil are constant. Similarly to
the water systems an equilibrium concentration can be also calculated by

the use of Henry's law (Dawson, 1977). The direction and the rate of ammonia flux is determined by the difference between the equilibrium and
the atmospheric concentrations. Since the direction of the flux was positive (deposition) in 53 cases of 62 measurements it can be assumed that
in most cases the atmospheric ammonia concentration exceeded the equilibrium one. The average concentrations were 2.66 and 1.93 $\mu g \cdot m^{-3}$ at upper and
lower level, respectively. The average difference between the concentrations at the two levels was 65 % while the average deposition rate was
0.017 $\mu g \cdot m^{-2} \cdot s^{-1}$ with a maximum of 0.102 and a minimum of -0.045
$\mu g \cdot m^{-2} \cdot s^{-1}$.

Because of the lack of equilibrium ammonia concentrations Δc was not
calculated. However, the deposition rates gave a correlation with the
concentration of ammonia measured at 4 m level:

$$d = 0.0297(NH_3) - 0.0619, \hspace{3cm} (7)$$

where d is the measured flux in $\mu g \cdot m^{-2} \cdot s^{-1}$, while the (NH_3) is the concentration measured at upper level in $\mu g \cdot m^{-3}$. The correlation coefficient is
r=0.49 at the probability level of 0.1 %. The somewhat poorer correlation
than for water surface suggests that the equilibrium ammonia concentration
(calculated from c_w and pH of the soil) should have been taken into account to obtain the Δc. It should be mentioned that the correlation coefficient between d and ammonia concentrations measured at lower level was
only r=0.06. It means that the atmospheric ammonia concentrations near
the surface are considerably affected by deposition or emission processes.

By using the formula concerning to the deposition velocity (v=d/c)
the values of v were also calculated for all cases. The average of 62
measurements was 0.48 $cm \cdot s^{-1}$ with a standard deviation of 1.1 $cm \cdot s^{-1}$. A
correlation was found between the deposition velocity and the reciprocal
value of ammonia concentration:

$$v = 1.04 - \frac{0.082}{(NH_3)}, \hspace{3cm} (8)$$

where v is the deposition velocity in $cm \cdot s^{-1}$, (NH_3) is the concentration
measured at upper level in $\mu g \cdot m^{-3}$. The correlation coefficient is 0.64 at
the probability level of 0.1 %. One can notice from equation (8) that
even at extremly high ambient concentrations the deposition velocity can
exceed the upper limit of 1 $cm \cdot s^{-1}$. It can also be seen from (8) that in
the case of v=0 the atmospheric ammonia level must be equal to equilibrium
one. This average equilibrium concentration is 0.85 $\mu g \cdot m^{-3}$. The lower limit of v cannot be calculated from (8) since in the case of low ammonia
concentrations the emission is controlled by diffusion processes in the
soil (Dawson, 1977).

As the main source of the atmospheric ammonia is the decomposition
of organics by bacterial activity the ambient ammonia concentration depends on the soil temperature (Georgii and Müller, 1974). In our cases
the atmospheric ammonia concentrations give correlation with the maximum
and minimum soil temperature (at 5 cm level). Correlation coefficients
are 0.54 and 0.57 at the probability level of 0.1 %.

5. CONCLUSION

The physical and chemical parameters of the natural waters and soils determine the equilibrium atmospheric ammonia concentration above these surfaces. Equilibrium concentration can be calculated by means of the solubility theory. The average value of this equilibrium level in the summer half-year is around 1 $\mu g \cdot m^{-3}$ over Hungary both in the cases of soil and water surfaces studied. Since the background average daily concentration is similar to the equilibrium value the sign of the difference between the atmospheric and equilibrium concentrations can be positive and negative. In other words deposition or emission can take place as a function of physical and chemical parameters of the air and water or soil. On the basis of these results the ammonia flux above all types of surfaces can be approximately calculated without measuring the flux. However, the knowledge of the temporal variation of the chemical composition both in waters and soils is necessary.

REFERENCES

Dawson, G.A.: 1977, J. Geophys. Res. 82, pp. 3125-3133.
Ferm, M.: 1979, Atmospheric Environment 13, pp. 1385-1393.
Georgii, H.W., and Müller, W.J.: 1974, Tellus 24, pp. 180-184.
Hales, J.M., and Drewes, D.R.: 1979, Atmospheric Environment 13, pp. 1133-1147.
Lau, N.C., and Charlson, R.J.: 1977, Atmospheric Environment 11, pp. 475-478.
Mészáros, Á., and Vissy, K.: 1974, J. Aerosol Sci. 5, pp. 101-110.
Scott, W.D., and Hobbs, P.V.: 1967, J. Atmosph. Sci. 24, pp. 54-57.

DRY DEPOSITION OF PARTICLES: COMPARISON OF PUBLISHED EXPERIMENTAL RESULTS WITH MODEL PREDICTIONS

E. Marggrander and D. Flothmann
Battelle-Institut e.V., Frankfurt am Main, Germany

ABSTRACT. Published particle depostion velocity data from well documented field and wind tunnel experiments are compared with the values predicted by Sehmel's semi-empirical model. It is found that at low shear stresses the predicted values are systematically greater than the measured ones. The increase of deposition velocity towards low shear stress predicted by Sehmel, being not explicable by known physical processes, has not been verified by the experimental data either. Systematic discrepancies between measured and predicted velocities as a function of particle size are also found.

CHARACTERIZATION OF THE USED MODEL

Dry deposition of aerosols is mainly determined by the following mechanisms:
- in the turbulent atmospheric layer by turbulent transport and sedimentation
- underneath in the laminar boundary layer by Brownian diffusion and sedimentation
- and eventually at the deposition surface by impaction and adhesion.

This height-dependent division is reflected in the model, developed by Sehmel and Hodgson [1]. Transfer through the two upper boxes, which is controlled by particle eddy diffusivity, Brownian diffusivity and settling velocity, is calculated from meteorological and aerosol variables. The remaining mass transfer resistance, which comprizes the influence of the surface is derived as a function of friction velocity u_* and of roughness length z_O. For the fitting procedure the authors have used their own wind tunnel deposition data obtained for five different surfaces (smooth brass shim stock, artificial grass, two sizes of gravel, water). Their experiment covered a wide range in the variables, namely:
- particle diameter $0.03\ \mu m \leqslant d \leqslant 29\ \mu m$
- friction velocity $0.11\ m/s \leqslant u \leqslant 1.44\ m/s$
- roughness height $0.001\ cm \leqslant z_O \leqslant 0.6\ cm$

H.-W. Georgii and J. Pankrath (eds.), Deposition of Atmospheric Pollutants, 23–30.

The model contains the following limitations and assumptions:
- surface elements should have small roughness heights only
- particle flux is constant
- particle eddy diffusivity can be determined
- the effect of gravity can be described
- particle agglomeration does not occur
- particles are completely retained by the surface.

Model predictions of Sehmel and Hodgson /1/ show that deposition
velocity
- decreases with increasing reference height
- increases with increasing roughness height
- in general increases with increasing friction velocity.

The latter is not true for particle sizes between 0.1 and 1 um at
small friction velocities. For this size range a minimum deposition
velocity occurs at a friction velocity between 0.2 and 0.3 m/s (see
fig 1). This minimum is not explained physically.

In order to test the validity of the model and its transferability to
field conditions, we tried to compare their model predictions with
experimental results by authors. The model results of Sehmel and
Hodgson /1/ were only available in graphical form at discrete friction
velocities u_* and roughness heights z_0. In order to obtain intermediate
values, we used a linear interpolation with respect to u_* and a
logarithmic one with respect to z_0.

EXPERIMENTAL DATA USED

In our comparison we restricted ourselves to publications on experiments, performed with monodispersed particles, and containing un-
processed data on deposition velocities as well as detailed information on particle characterics (size, density), meteorological and
surface conditions. Table 1 gives a summary of the experiments selected
and of the particle sizes, surfaces, roughness lengths and friction
velocities covered.

Table 1 further indicates if the experiments were performed in the
field or in a wind tunnel. One sees that in field experiments preferably a single particle size or only a small size range has been
investigated at different meteorological conditions whereas in wind
tunnel experiments the deposition velocity of different particle sizes
can be measured at fixed friction velocities.

RESULTS OF THE COMPARISON

In figure 2a) the results by Chamberlain /2/ on deposition velocities
of Lycopodium spores (32 um diameter), measured in a w i n d -
t u n n e l, are compared with the corresponding values predicted by

Table 1: Summary of the experiments selected for comparison

Author	Particle diameter μm	Surface	z_O cm	u_* m/s	Type of experiment
Chamberlain /2/	32	grass grass+soil Petri dish	0.4-7.5	0.12-0.94	field
Chamberlain /2/	32	grass artificial grass towelling rough glass	0.6 1.0 0.045 0.02	0.18-1.83 0.28-1.46 0.58-1.26 0.17-1.25	wind tunnel
Chamberlain /2/	32,19, 5, 2, 1, 0.08	grass artificial grass	0.6 1.0	0.35, 0.7 1.4	wind tunnel
Clough /3/	32, 3.5-6.5 ~3	moss	0.37	0.37,0.87	wind tunnel
Jonas /4/	2.0-5.7	grass		0.06-0.32	field

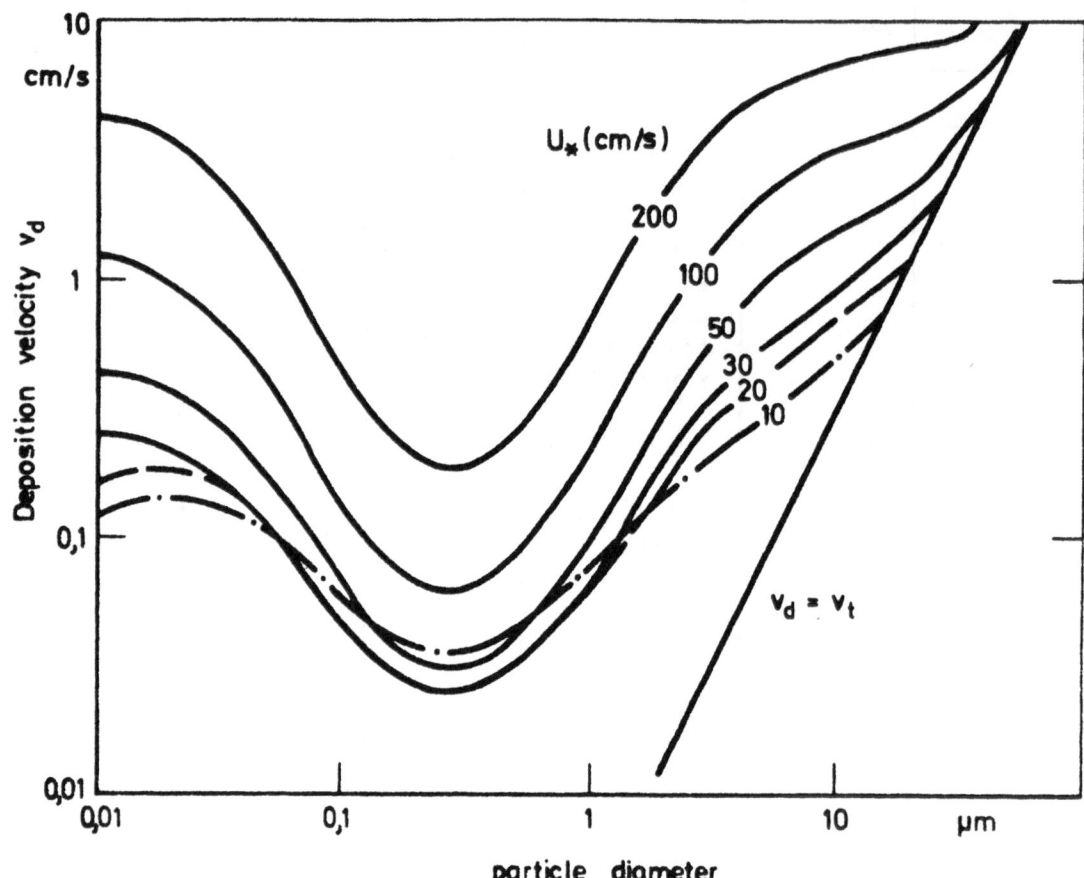

Fig. 1: Deposition velocity as a function of particle
   diameter and friction velocity u_* , predicted
   by Sehmel and Hodgson /1/.
   Reference height z=1 m, roughness height z_o=3.0 cm.
   neutral stability.

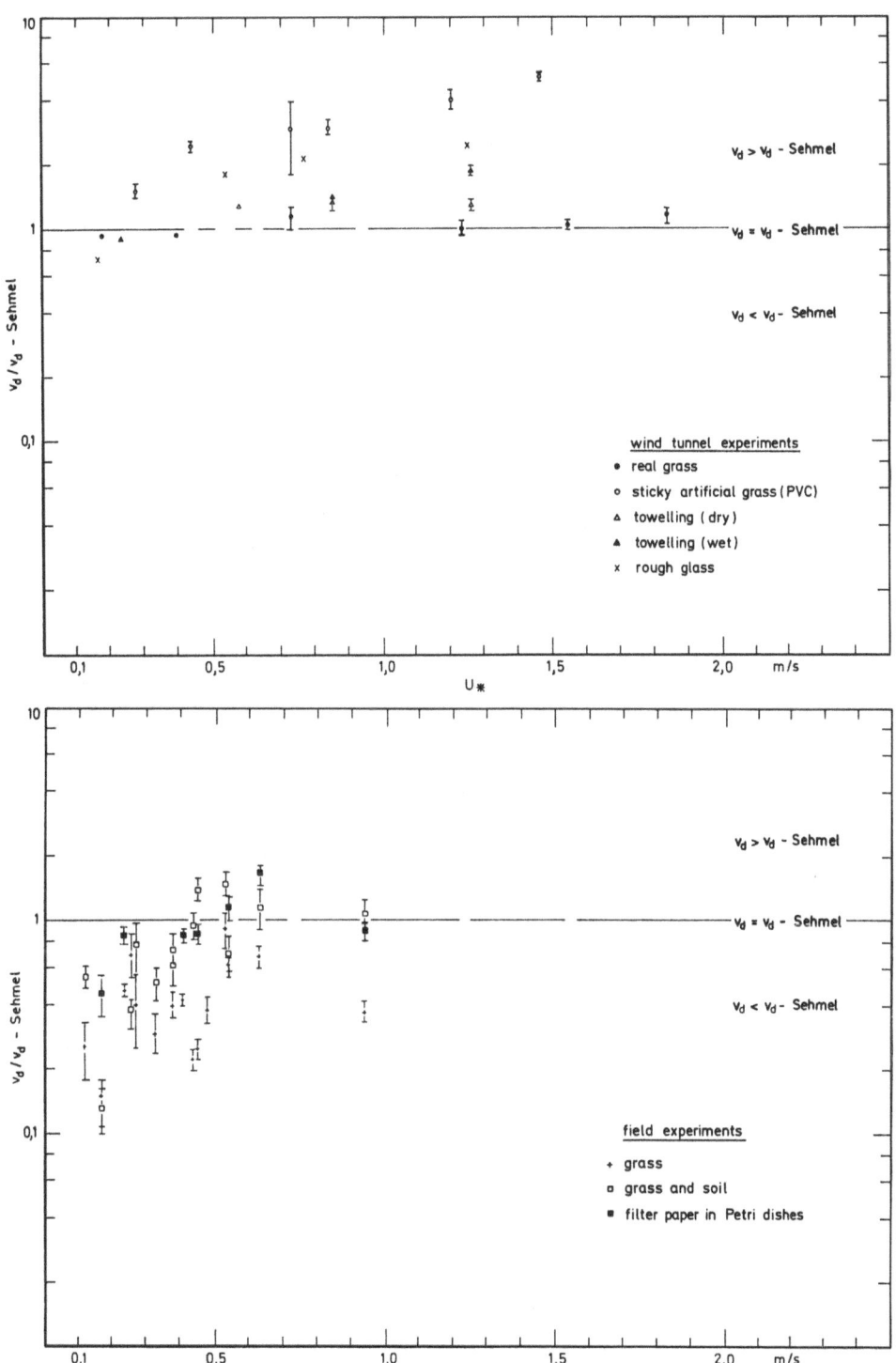

Fig. 2: Ratio of the experimental to the predicted value of the
deposition velocity as a function of friction velocity u_*.
a) Wind tunnel experiments
b) Field experiments
with Lycopodium spores (32 µm diameter) by Chamberlain /2/.

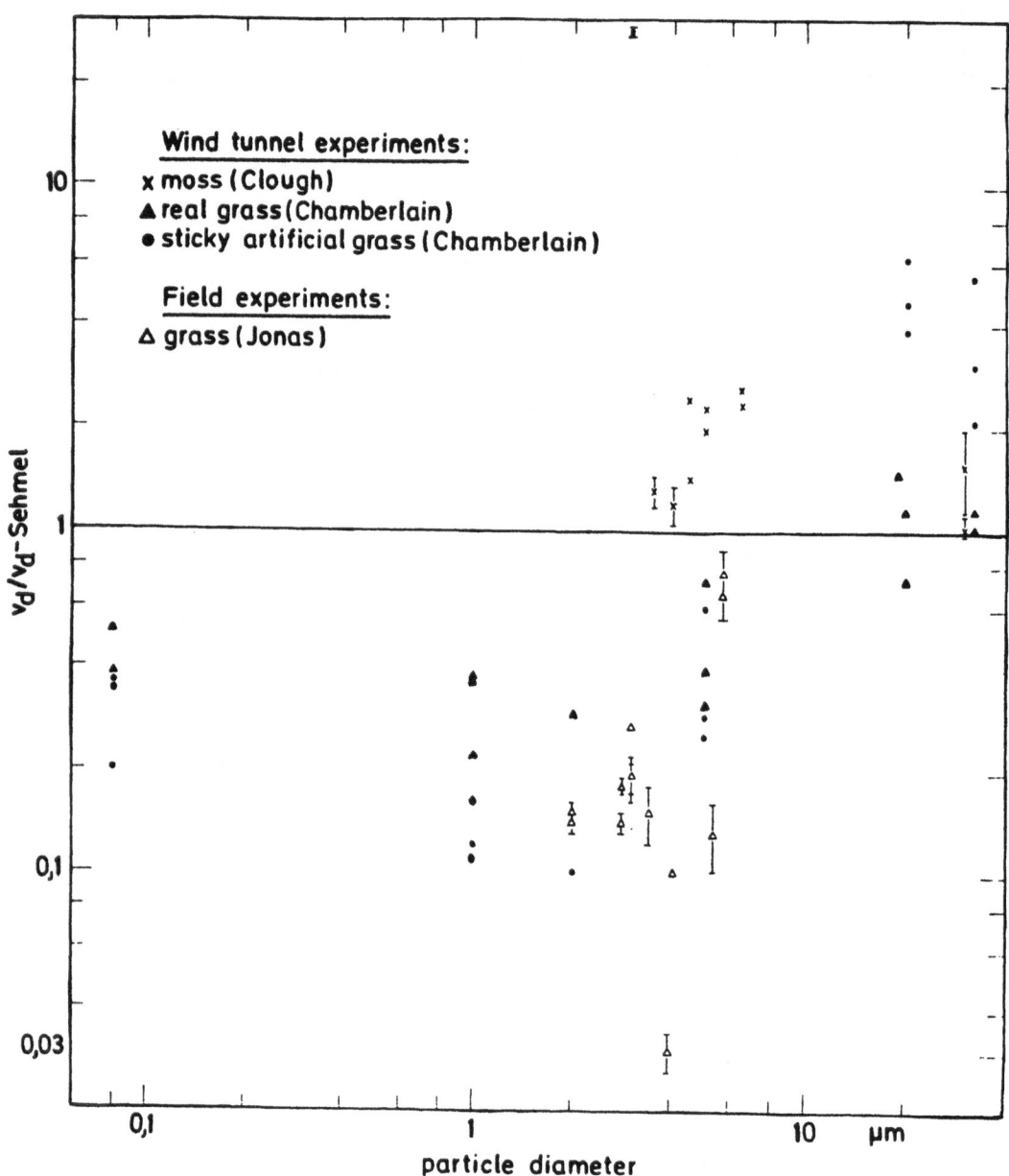

Fig. 3: Ratio of the experimental to the predicted value of
the deposition velocity as a function of particle size.
Wind tunnel experiments by Chamberlain /2/ and Clough /3/
and field experiments by Jonas /4/.

Sehmel and Hodgson. The ratio of the experimental to the predicted
value of the deposition velocity is plotted as a function of friction
velocity u_*. For the experiment with real grass an excellent agreement
is obtained. For artificial grass, however, (as well as for other
artificial structures) at increasing friction velocity the wind tunnel
results are increasingly higher than prediction, although Sehmel and
Hodgson have used (own) experiments on artificial grass and other
artificial structures in their fitting procedure.
The results of Chamberlain /2/, obtained with Lycopodium spores in
f i e l d   e x p e r i m e n t s, are also compared with the model
prediction in fig.2b).In the most cases the experimental values, which
were all measured at small friction velocities, are lower than the predicted deposition velocities. The smallest deviations between experiment and model are found for deposition onto "grass and soil".
When combining wind tunnel and field results, one recognizes a distinct
dependency of the deviations on the friction velocity u_*. At small u_*
the experimental values are tending to be lower and at large u_* to be
higher than prediction.
Fig. 3 shows a comparison of deposition velocities as a function of
particle size. Except the experimental results of Jonas /4/ the data
are derived from wind tunnel experiments. One finds that for large
particle sizes experimental values are larger and for small particles
smaller than the predicted values. Since the data of Chamberlain /2/
plotted in fig. 3 are all measured at a fixed set of u_* values, one
can conclude that this trend due to particle size is independent of u_*.

SUMMARY

Discussing wind tunnel and field experiments separately, one can draw
the following conclusions:
The comparison shows, that the f i e l d   e x p e r i m e n t s
selected, having preferably been made at low friction velocities, lead
to smaller deposition velocities than those predicted by the model of
Sehmel and Hodgson. This may be attributed to different reasons:
- The field experiments may contain systematic errors.
- The model may contain assumption not being met by field experiments.
- The model was developped by fitting wind tunnel experiments,
  representing conditions not transferable to field situations.

Results from w i n d   t u n n e l experiments also show systematic
deviations from the model prediction:
- At high friction velocities experimental velocities are
  larger than predicted ones. The deviation depends on the quality of
  surface.
- For small particle sizes experimental deposition velocities
  are smaller, for large sizes larger than predicted ones.

Deviations of deposition velocities are high resp. low up to one
order of magnitude. The systematic deviations depending on both,
particle size and friction velocity, suggest the deviations to be
caused by varying efficiencies in the impaction process.

ACKNOWLEDGEMENT

We would like to thank Dr. D. Wagenbach, Institut for Environmental
Physics of the University of Heidelberg, for valuable comments and
discussions.
This contribution is part of the work, sponsored by the German
Federal Environmental Agency (UBA), Berlin under contract
no. 104 02 609.

REFERENCES

1.   G.A. Sehmel und W.H. Hodgson: A model for predicting dry
     deposition of particles and gases to environmental surfaces.
     PNL-SA-6721-Rev. 1.

2.   A.G. Chamberlain: Transport of Lycopodium spores and other
     small particles to rough surfaces. Proc. R. Soc., 296: 45-70
     (1966).

3.   W.S. Clough: The deposition of particles on moss and grass
     surfaces. Atm. Env. $\underline{9}$: 1113-1119 (1975).

4.   R. Jonas: Statusbericht über die Feldversuche zur Bestimmung
     der Ablagerungsgeschwindigkeit von Aerosolen. KFA-Jülich,
     ZST-Bericht Nr. 0295, (1979).

DRY DEPOSITION VELOCITIES OF AEROSOL SULPHATE IN RURAL EASTERN ENGLAND

T.D. Davies and K.W. Nicholson
School of Environmental Sciences, University of East Anglia,
Norwich, U.K.

Dry deposition velocities of aerosol sulphate have been determined by the gradient method, using non-dispersive X-ray flourescence. The measurements were made over periods of several hours, during relatively constant atmospheric conditions, day and night, throughout one year. The median deposition velocity for the whole period was 0.08 cm s^{-1}, a value close to earlier experimental and many theoretical determinations, but considerably less than some recent, short-period, field determinations. Although there is considerable scatter, the deposition velocity varies with atmospheric stability and with the seasons. The dry deposition of aerosol sulphate is much less important than the dry deposition of sulphur dioxide and the wet deposition of sulphate, but is of the same order as the wet deposition of sulphur dioxide.

1. INTRODUCTION

The relative importance of dry and wet deposition of sulphur has been the subject of recent interest (e.g. OECD, 1977; Garland, 1978). Sulphur dioxide and particulate sulphate comprise the predominant forms of atmospheric sulphur in the developed countries (Garland, 1978). Wet removal rates of sulphur are generally well-known although the contribution of rainborne SO_2 to precipitation sulphur is known with less confidence (Davies, 1976, 1979a; Hales and Dana, 1979; Dana, 1980). Of the dry deposition contributions to sulphur removal from the atmosphere, more work has been done on the sulphur dioxide component because of the relatively small particulate sulphate dry deposition rate (Garland, 1978) and greater experimental difficulties (Sehmel, 1980).

2. DRY DEPOSITION OF PARTICLES

The surface dry deposition rate of a pollutant is usually expressed in terms of the deposition velocity, v_d (Chamberlain and Chadwick, 1953) where the flux from the atmosphere to the ground is estimated as the product of the atmospheric concentration and the deposition velocity

H.-W. Georgii and J. Pankrath (eds.), Deposition of Atmospheric Pollutants, 31–42.

for a particular height (usually 1.0 m over land). For particle diameter sizes of less than about 0.1 µm, v_d is influenced by Brownian
diffusion and for sizes greater than about 1.0 µm diameter, vd is
accelerated by gravity settling. Only in the size range 0.1 µm -
1.0 µm diameter is v_d nearly constant for a selected roughness length,
particle density and wind speed, and eddy diffusion is regarded as the
important deposition process (Sehmel, 1980).

Studies of the size distribution of sulphate aerosol indicate
that by far the greater proportion of the mass falls in the size-range
0.1 to 1.0 µm diameter (e.g. Heard and Whiffen, 1969; Georgii et. al.,
1971; Mészáros, 1970). Cawse (1974) and Mészáros (1978) show that more
than 90% of particulate sulphate mass is in this size range in rural
aerosol. A minor fraction of the sulphate mass is sometimes found
between 1.0 µm and 20.0 µm and this might contain sea-salt sulphate and
sulphate in wind-blown soil (Garland, 1978).

Wind tunnel studies shows that the v_d of 0.1 - 1.0 µm particles
over relatively smooth surfaces are very small, from less than 0.01 to
about 0.05 cm s^{-1} (Chamberlain, 1967; Sehmel, 1973; Clough, 1973;
Craig et. al., 1976). Wind tunnel work by Clough (1973) suggested v_d
for 0.5 µm particles on grass (roughness length, z_0, about 1.0 cm)
varied linearly from 0.01 to 0.05 cm s^{-1}, with friction velocity (u_*).

Sehmel (1980) reviewed previous field experiments and pointed out
the large range in the values of v_d. He stressed that field experiments
require very good control and definition of all experimental conditions.
The results from these field experiments have not been generalised to
within an order of magnitude because of experimental uncertainties and
limited data. A major uncertainty in field experiments which employ
some sort of collector is the relationship between deposition on the
collector and deposition on the natural surface. Studies which have
measured radioactive particle or trace metal deposition on collectors
(e.g. Peirson and Cambray, 1965; Pierson et. al., 1973) find v_d in the
range 0.05 - 0.30 cm s^{-1} (for particles which are, presumably, less
than 1.0 µm diameter).

From theoretical considerations, Sehmel (1980), elaborating on
Sehmel and Hodgson (1978), showed that for a surface with a z_0 of 10 cm
and u_* of 30 cm s^{-1}, the v_d for particles in the range 0.1 - 1.0 µm
diameter would be around 0.05 cm s^{-1}. Slinn (1976) indicated that the
v_d of 0.1 - 1.0 m particles over relatively smooth surfaces is very
small (around 0.01 cm s^{-1}).

Most of the experimental and theoretical work indicates that the v_d
of sulphate particles (the greater proportion of which is assumed to be
in the range 0.1 - 1.0 µm diameter) over grass does not exceed around
0.1 cm s^{-1} (Garland, 1978). The v_d adopted by the OECD (1977) on a
regional scale was 0.2 cm s^{-1}. Convair (1960) found that field deposition velocities increased with wind speed, although they were independent of atmospheric stability. Sehmel and Hodgson (1978) predict that

v_d is nearly independent of atmospheric stability.

Recent field studies have suggested rather higher v_d values for particles in the range 0.1 - 1.0 μm. Wesely et. al. (1977) contend that, "for particles smaller than 1.0 μm, the value of v_d in moderate to light winds over many natural surfaces may be about 1.0 cm s^{-1}, rather than the value 0.1 cm s^{-1} that is currently popular with numerical modelers".The authors also point out that atmospheric stability has a dominant role in controlling v_d, which implies that diurnal variations ought to be taken into account. This study was based on a small number of observations made over a couple of hours.

Everett et. al. (1979) also report v_d values substantially greater than those in most contemporary numerical models. From measurements made over an eight-day period at heights of 11.5 m and 34.5 m on a tower, they found a mean v_d of 1.4 cm s^{-1} for particulate sulphur, and a strong dependence on atmospheric stability. Sheih et. al. (1979) have estimated v_d for particulate sulphate over the Eastern United States and surrounding regions. They point out that v_d will depend on surface characteristics and atmospheric conditions. The surface resistance to particulate deposition which is used by Sheih et. al. is derived from recent experimental work by such workers as Wesely et. al. (1977). The computed particulate sulphur v_d values are correspondingly comparable with the results of Weseley et. al..

It is evident that there is a considerable discrepancy between v_d values estimated from laboratory studies, most theoretical work, earlier field studies and the more recent and more sophisticated field experiments. Experimental conditions in the laboratory and in the field will differ; it is impossible to limit the particulate size-range in the field, and the atmosphere is more variable than the wind-tunnel atmosphere. The recent field experiments summarised here have been conducted over very short periods.

3.   MEASUREMENT OF SULPHATE DRY DEPOSITION

At a rural site in eastern England remote from major sources of sulphur, except the small city of Norwich 16 km to the east (further details in Davies, 1979a), the wet and dry deposition of sulphur dioxide and particulate sulphate have been measured, on a semi-continuous basis, over an observation period of some years (Davies, 1979a, 1979b). The location of the rural observation station is shown in figure 1. The wet deposition of sulphur dioxide and sulphate was not measured contemporaneously with the dry deposition rates of sulphur dioxide nor with the dry deposition rates of particulate sulphate, but all fluxes were measured over a long enough period for realistic estimates to be made of the components of total sulphur deposition to the ground at this site.

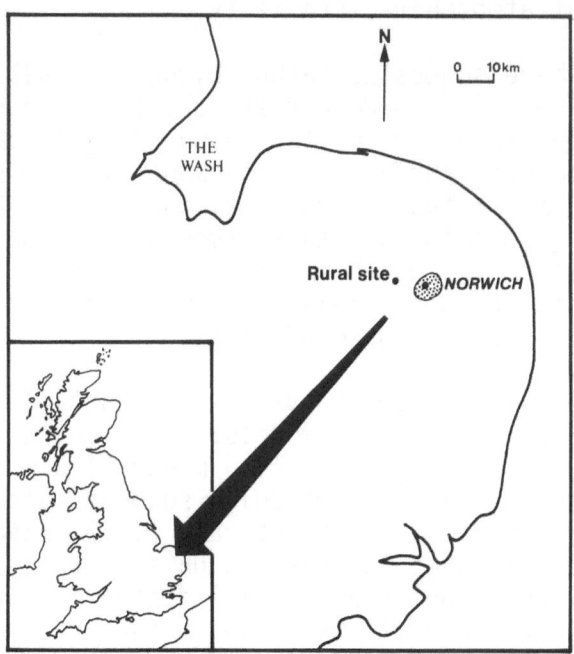

Figure 1. Location of the observation station.

The observations made of the dry deposition rates of particulate sulphate are reported in detail here. Figure 2 shows the local geography of the sampling site; the local relief was flat and the fetch

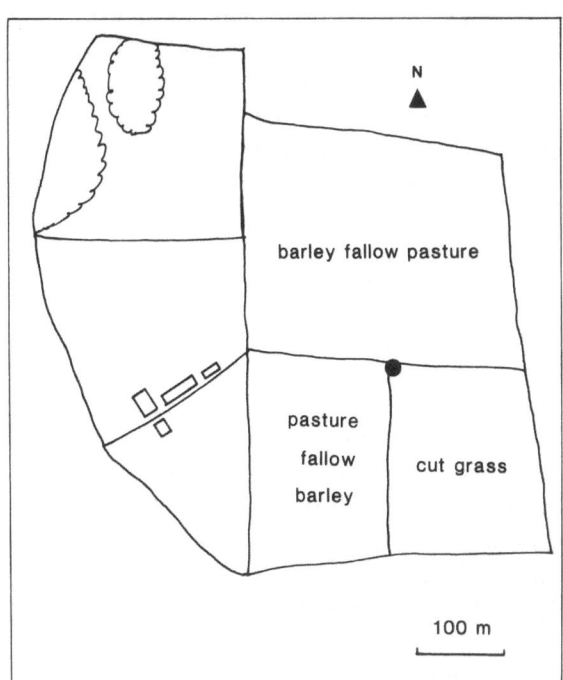

Figure 2. Details of the sampling station. The nature of each field-surface throughout the sampling year (June-June) is indicated.

was adequate in all directions. The sampling equipment was installed near the boundary of three fields (figure 2): 1) for logistical reasons; to avoid conflict with agricultural requirements and because the investigators had no control over the crop regime: 2) to achieve a "representative" medium-term assessment of dry deposition for rural eastern England: 3) in order to study the relationship between dry deposition and differing surfaces and their related atmospheric flow. The sampling equipment was sited to avoid "edge effects", i.e. right at the boundary.

Air was aspirated through filter holders which housed 47 mm diameter cellulose acetate (Oxoid) membrane filters. Once the air had passed through the filter it bubbled through a dreschel bottle containing hydrogen peroxide in order to remove sulphur dioxide. Small annular heaters were fitted around the samplers so that the temperature of the incoming air was raised by around $10^{\circ}C$ above ambient temperature in order to avoid take-up of sulphur dioxide on the filter or particulate material. The filter holders were designed so as to disturb the airflow as little as possible. The filters are surface collectors which act as a sieve and impactor for particles > 0.3 μm, particles < 0.3 μm are collected through diffusion. The collection efficiency is around 99.9%.

The filter holders were fixed in position on a tubular mast at five heights from 0.23 m to 2.30 m. For each sampling run, the filter holders were adjusted to face into the wind. This precaution was regarded as very important since simultaneous runs of five samples on two masts, one pointing into the wind and the other facing away from the wind, yielded average particulate sulphate concentration depressions of up to 50% for the samplers facing away from the wind (for the adopted aspiration rate of 5.0 1 min^{-1}). The total volume of air aspirated for each sampler was recorded by dry gas-meters. The sulphate concentration on the filter was measured by non-dispersive X-ray flourescence (NDXRF). The attenuation of the flourescent X-rays by the filter medium is only about $+2.5$%, since the medium is essentially a surface filter. The sulphate standards were made through forced filtration of a uniformly suspended barium sulphate precipitate onto blank filters. The precipitate was prepared by the reaction of sodium sulphate solution with barium chloride solution. In order to minimise the amount of sulphate in solution, a relatively weak solution of sodium sulphate was added to the barium chloride. Analytical error was minimised by using the same batches of filters and the total analytical accuracy was estimated at $+$ 0.25 μg m^{-3}. NDXRF determines sulphur and so, should any organic sulphur be present on the filter, it would register an elevated sulphate level. It is assumed that the NDXRF-determined sulphur represents particulate sulphate on the filter. On some occasions when it was suspected that soil was being blown off a field (often leading to apparent upward fluxes) the data were not used.

Air temperature was measured at six heights by thermistors ($+0.1^{\circ}C$) and wind speed was measured at five heights by reed-switch anemometers with a threshold of 0.3 m s^{-1} (accuracy $+5$%).

The gradient method (Atkins and Garland, 1974) was employed to determine the flux of particulate sulphur to the ground. Corrections for stable conditions (Richardson number, R_i, > 0.01) and unstable conditions (R_i < -0.01) were made by utilizing the dimensionless shears of momentum and heat determined from their empirical relationships with R_i (the equations supported by Thom et. al., 1975). The deposition velocities, v_d, were then calculated with reference to the height 1.0 m.

Observations were made throughout the period June 1979 - June 1980. The average sampling period was five hours, and the sampling was restricted to periods when atmospheric conditions were relatively constant. The gradient method of flux measurement requires that airborne concentrations and atmospheric variables should be more or less constant. So the sampling period never, for example, spanned the dusk or the dawn period. The distribution of the sampling runs is shown in figure 3. About equal numbers of fluxes were determined for day and for night. So the novelty of the study reported here is the medium-term nature of the observations made over a land surface "representative" of eastern England, under a variety of weather conditions. The crops grown in the fields throughout the period are indicated in figure 2. The south-east field always consisted of closely-cut grass. The changing crop heights are incorporated into the flux determinations via z_0, the displacement height d, and u_*. It is estimated that the accuracy in the concentration gradients is around ±30% and the accuracy in v_d is around ±40% for average airborne concentrations and gradients. When the concentration gradient and/or the concentration is very small, the error may be considerably greater.

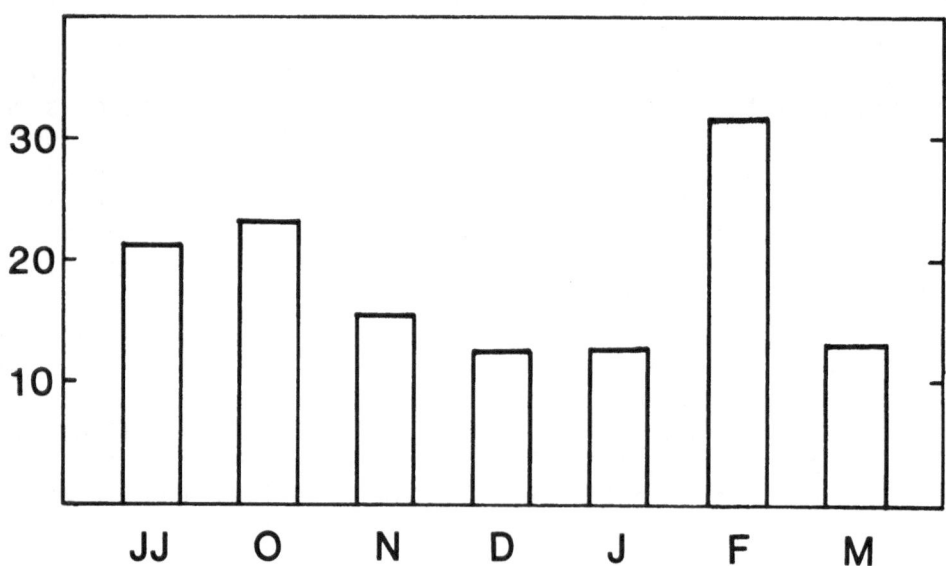

Figure 3. Frequency distribution of v_d determinations throughout the observation period (JJ, June and July; M, March).

4. THE OBSERVED DEPOSITION VELOCITIES

Figure 4 shows the distribution of calculated v_d of particulate sulphate over the period of one year. The negative values are for cases

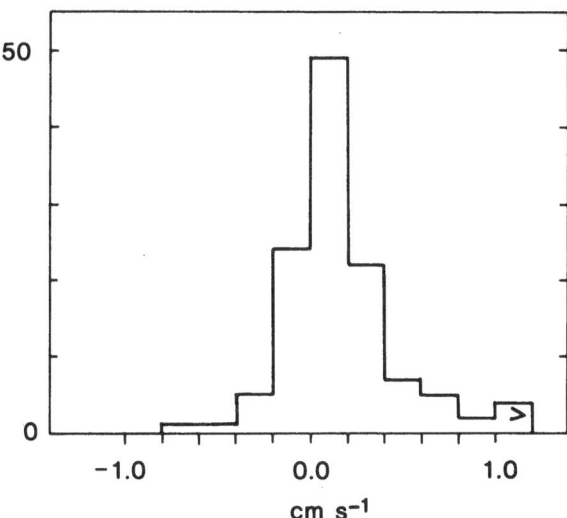

Figure 4. Frequency distribution of v_d values throughout the observation period. The classes are (±) 0.00-0.19, 0.20-0.39, 0.40-0.59, 0.60-0.79, 0.80-0.99, ⩾ 1.00 ($\overline{cm}$ s^{-1}).

where the measurement errors could put a positive sign on the value, or where the concentration gradients are small and close to, or lower than, the probable limit of detection ($\sim$ 0.10 μg m^{-3} over the 2.0 m height). Some of the positive v_d values also incorporate this uncertainty but it is assumed that the descriptive statistics of the total population have some physical validity. The median value is 0.08 cm s^{-1}; the median value is probably more appropriate than the arithmetic mean (0.23 cm s^{-1}) because of the heavy weighting of extremes. Discounting the 'extremes' (v_d > 1.0 cm s^{-1}) the distribution approximates to a normal distribution.

Figure 5 shows the distribution of v_d by "stability class". The median and mean v_d values for the different stability classes are: neutral, 0.04 cm s^{-1}; stable, 0.025 cm s^{-1}; unstable, 0.16 cm s^{-1}. The respective arithmetic means are: 0.22, 0.01 and 0.64 cm s^{-1}. The shapes of the distributions are interesting; the neutral case is close to a normal distribution, the stable case has a negative skew, and the unstable case is positively skewed. The stability correction (described in the previous section) would tend to stretch the stable and unstable distributions anyway (if the appropriateness of the correction is questioned, then any shift in distributions might be regarded simply as a function of the correction applied) but would not affect the skewness of the distributions. These results tend to confirm the dependence of v_d on atmospheric stability found by Wesely et. al. (1977) and

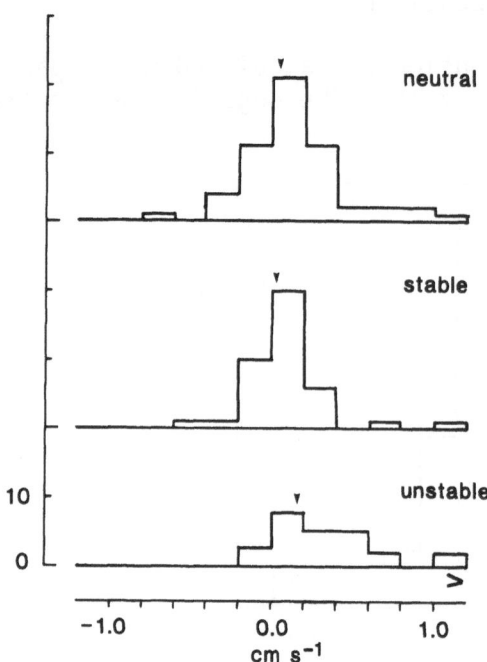

Figure 5. Frequency distribution of v_d by stability class. Neutral, R_i, -0.01 to +0.01; stable, > +0.01; unstable, < -0.01. The arrows represent the median values of each distribution (see text).

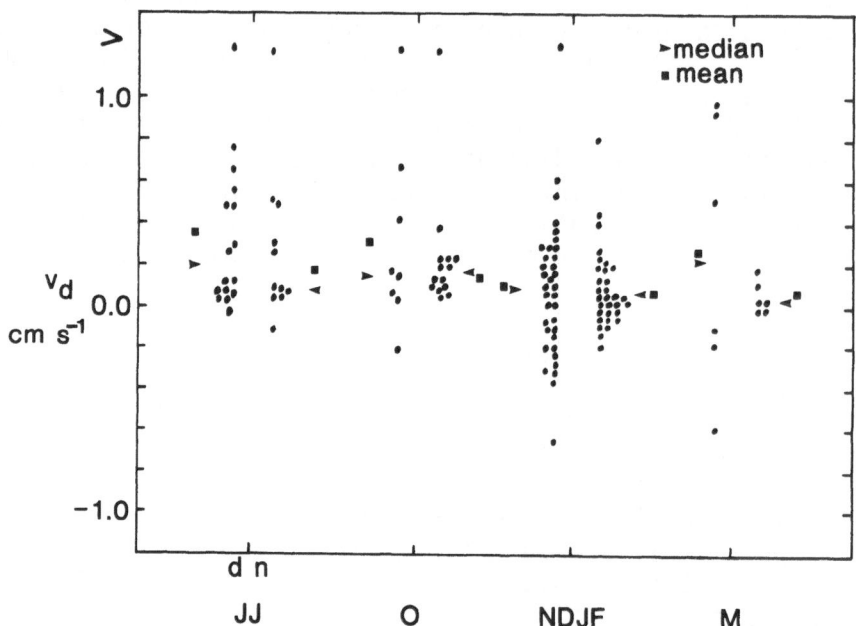

Figure 6. Distribution of v_d values by month (June + July, October, November + December + January + February, March) and by day (d) and night (n).

Everett et. al. (1979). The relationship between v_d and R_i (on an individual observation basis) is negative but is not statistically significant; other factors produce too much scatter.

Figure 6 illustrates the seasonal variation in particulate sulphate v_d. The division of the v_d values into the particular groups of months is based on number of available observations as much as anything. On the basis of the observations shown here, there does appear to be an annual cycle. The median values for June + July are 0.20 (day) and 0.08 (night) cm s^{-1}; for November + December + January + February the median values are 0.08 (day) and 0.05 (night) cm s^{-1}. The atmosphere is more stable at night and negative lapse rates are stronger during the summer day than the winter day.

Besides thermal stability, wind speed and surface aerodynamic roughness affect v_d. Roughness length is greater in the summer, unless the crop is mown. There is a weak (statistically significant) positive relationship between v_d and z_0. The considerable scatter may be due, in part, to the problem of negative v_d values. Figure 7 illustrates the distribution of z_0 throughout the year. The large ranges indicate

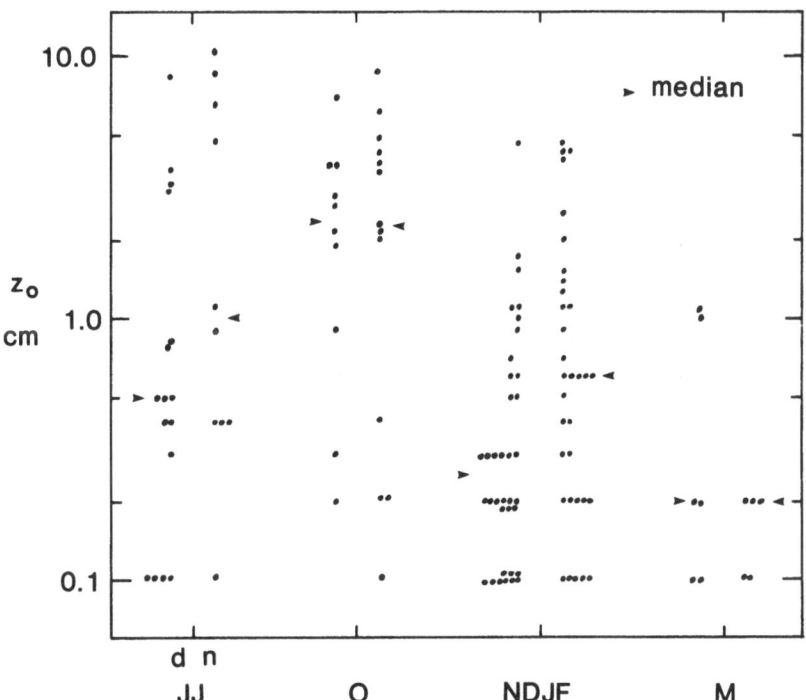

Figure 7. Distribution of roughness lengths by month and by day (d) and night (n).

differing fetches (function of crop height and wind direction), but the summer z_0 values are seen to be higher than the winter values. The weak correlation between v_d and z_0 on a single observation basis may also be related to evidence of z_0 decreasing with increasing wind speed as the crop "bends over" (also pointed out in Sehmel, 1980).

The friction velocity, u_*, also exhibits a seasonal variation (figure 8) for the observation periods (u_* is a function of z_0). On an individual observation basis, using all the v_d values, there is a weak positive correlation between log (modulus v_d value) and u_* ($r = 0.4$, significant at the 1% level), although the nature of the relationship is difficult to interpret because of the inclusion of the negative

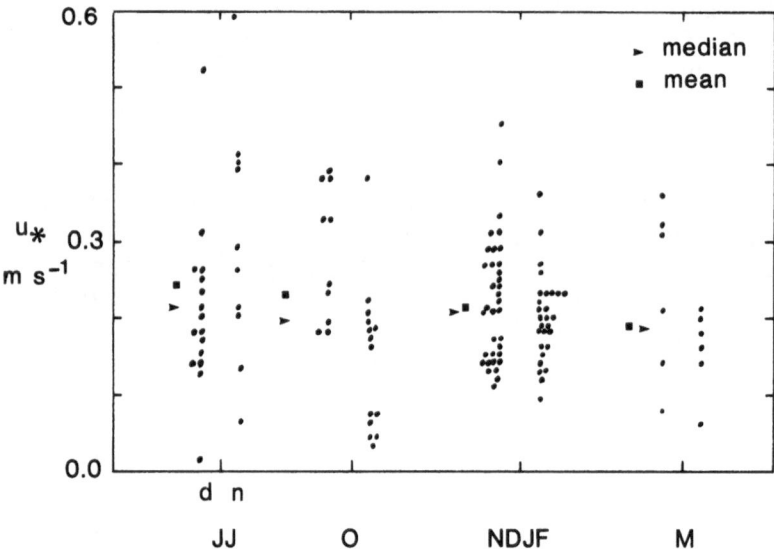

Figure 8. Distribution of u_* by month and by day and night.

v_d values. Excluding the negative v_d values yields a relationship of $\log_{10} (v_d) = 2.7 u_* - 1.4$ ($r = 0.41$, significant at the 1% level). The slope of this regression is steeper than in most previous experimental or theoretical work, but the results are not altogether directly comparable because z_0 is also varying here.

Although all the wet and dry deposition observations made in this study were not wholly contemporaneous, they were made over long, and representative, enough periods for a sulphur deposition budget to be estimated (table 1). The observations of the wet deposition are described in Davies (1979a); the sulphur dioxide dry deposition rates are still to be reported. These values exhibit reasonable agreement with

	Dry	Wet
SO_2	2.70	0.13
SO_4^{2-}	0.11 (winter 0.03)	0.79
	(summer 0.08)	

Table 1. Annual sulphur deposition, g (S) $m^{-2} y^{-1}$ over a surface "representative" of agricultural eastern England.

other estimates for eastern England (e.g. OECD, 1977). Dry deposition of particulate sulphate is seen to be of considerably less importance than the dry deposition of sulphur dioxide. The 'summer' dry deposition rate of particulate sulphate is considerably greater than the Winter' deposition rate.

5. CONCLUSIONS

The deposition velocities of particulate sulphate, determined from observation periods of a few hours (during relatively constant atmospheric conditions) throughout a period of one year, are of the same order as, or slightly higher than, the v_d values determined from theoretical and all but the most recent experimental work. The more recent field experimental work, based on relatively short observation periods, indicates deposition velocities of an order of magnitude higher. This study confirms the recent experimental work which suggests that the deposition velocity is dependent upon atmospheric stability. The deposition velocities exhibit weak associations with friction velocity and roughness length but the scatter is considerable; an indication that real atmospheric and surface conditions are very variable. This point is an important one to bear in mind when considering experimental determinations of v_d. It was for this reason that the v_d of particulate sulphate was measured over a long period, under a variety of surface and atmospheric conditions, in order to obtain a reliable estimate of the "characteristic" deposition velocity. The measurements indicate that of the four main components of sulphur deposition in rural eastern England, the dry deposition rate of particulate sulphate is similar to the wet deposition of sulphur dioxide and considerably less than the dry deposition of sulphur dioxide or the wet deposition of sulphate. The suggestion by Garland (1978) that the deposition velocity for sulphate aerosol is "no larger than 0.1 cm s^{-1}" seems to be substantiated in terms of "average" values, but changing atmospheric and surface conditions produce a considerable range of values.

Acknowledgements: this study formed part of a project funded by the Natural Environment Research Council.

REFERENCES

Atkins, D.H.F. & Garland, J.A. 1974, Observation and Measurement of Atmospheric Pollution, WMO-No. 368, Geneva, pp. 579-594.
Cawse, P.A. 1974, A survey of atmospheric trace elements in the UK (1972-73), AERE-R 7669, HMSO, London.
Chamberlain, A.C. 1967, Proc. R. Soc. Lond. A, 296, pp. 45-70.
Chamberlain, A.C. & Chadwick, R.C. 1953, Nucleonics 8, pp. 22-25.
Clough, W.S. 1973, Aerosol Sc. 4, pp. 227-234.
Convair. 1960, Fission products field release test II, NARF 60-10T

(FZK-9-149; AFSWC-TR-60-26), US Airforce Nuclear Aircraft Research
Facility (availableNTIS, Dept. of Commerce, Springfield, VA, USA).
Craig, D.K., Klepper, B.L., & Buschbom, R.L. 1976, Proceedings Atmos-
phere-Surface Exchange of Particulate and Gaseous Pollutants (AEC
Symp. Series 740921, NTIS) pp. 224-261.
Dana, M.T. 1980, Journ. Geophys. Res., 185, no. C8, pp. 4475-4480.
Davies, T.D. 1976, Atmos. Environment, 10, pp. 879-890.
Davies, T.D. 1979a, Atmos. Environment, 13, pp. 1275-1285.
Davies, T.D. 1979b, Sulphur Emissions and the Environment, The Society
of Chemical Industry, London, pp. 212-214.
Everett, R.G., Hicks, B.B., Berg, W.W. and Winchester, J.W. 1979,
Atmos. Environment, 13, pp. 931-934.
Garland, J.A. 1978, Atmos. Environment, 12, pp. 349-362.
Georgii, H-W., Jost, D. and Vitze, W. 1971, Konzentration und
Grössenverteilung des Sulphataerosols in der unteren and mitteleren
Troposphäre. Berichte des Institutes für Meteorologie und Geophysik
der Universität Frankfurt/Main, No. 23.
Hales, J.M. & Dana, M.T. 1979, Atmos. Environment, 13, pp. 1121-1132.
Heard, M.J. & Whiffen, R.D. 1969, Atmos. Environment, 3, pp. 337-340.
Mészáros, E. 1970, Tellus, 22, pp. 235-238.
Mészáros, E. 1978, Atmos. Environment, 12, pp. 2425-2428.
OECD, 1977, Long Range Transport of Air Pollutants, Measurements and
Findings, OECD, Paris.
Sehmel, G.A. 1973, Aerosol Sci., 4, pp. 125-138.
Sehmel, G.A.1980, Atmos. Environment, 14, pp. 983-1011.
Sehmel, G.A. & Hodgson, W.J. 1978, A Model for Predicting Dry Deposi-
tion of Particles and Gases to Environmental Surfaces, PNL-SA-6721,
Battelle, Pacific Northwest Laboratory, Richland, WA, USA.
Shieh, C.M., Wesely, M.L. & Hicks, B.B. 1979, Atmos. Environment, 13,
pp. 1361-1368.
Slinn, W.G.N. 1976, Proceedings Atmosphere-Surface Exchange of Parti-
culate and Gaseous Pollutants (AEC-Symp. Series 740921, NTIS),
pp. 1-40.
Thom, A.S., Stewart, J.B., Oliver, H.R. & Gash, J.H.C. 1975, Quart.
Journ. Roy. Meteorol. Soc., 101, pp. 93-105.
Peirson, D.H. & Cambray, R.S. 1965, Nature, 205, pp. 433-440.
Peirson, D.H., Cawse, P.A., Salmon, L. & Cambray, R.S. 1973, Nature,
241, pp. 252-256.
Wesely, M.L., Hicks, B.B., Dannevik, W.P., Frisella, S. & Husar,
R.B. 1977, Atmos. Environment, 11. pp. 561-563.

RESIDENCE TIME AND DEPOSITION OF PARTICLE-BOUND ATMOSPHERIC SUBSTANCES

Jürgen Müller
Umweltbundesamt - Pilotstation Frankfurt
Feldbergstr. 45, Frankfurt/M.

ABSTRACT

By aid of measured mass size distributions the atmospheric residence times for particle-bound substances can be determined. The distributions are considerably barying and are related to the vapour pressure of the substance. By assumption of a stationary equilibrium in the airborne state the residence times can be related to each other by use of the boiling and melting points. The absolute values are mainly influenced by the wet residence time - determined by the atmospheric water cycle - of the considered region.

The residence time within the mixing layer of the troposphere is inversely proportional to the total deposition velocity. According to their mass size distributions the substances can be classified into local, regional and long-range pollutants. An estimate of the dry to wet deposition ratio of a substance can be made. The behaviour of metals and the acid rain ions $SO_4^=$, NO_3^- and Cl^- is discussed.

1. INTRODUCTION

The fate of substances injected into the atmosphere is determined by the meteorological conditions and their individual properties. In order to describe the different behaviour of the substances the atmospheric quantities should also be related to the physico-chemical parameters.

The residence time is an important quantity for atmospheric description. It is defined as the mean time of residence for a molecule of a substance injected into the atmosphere. The residence time gives us a measure how fast the atmosphere is able to digest the substance.

43

H.-W. Georgii and J. Pankrath (eds.), Deposition of Atmospheric Pollutants, 43–52.

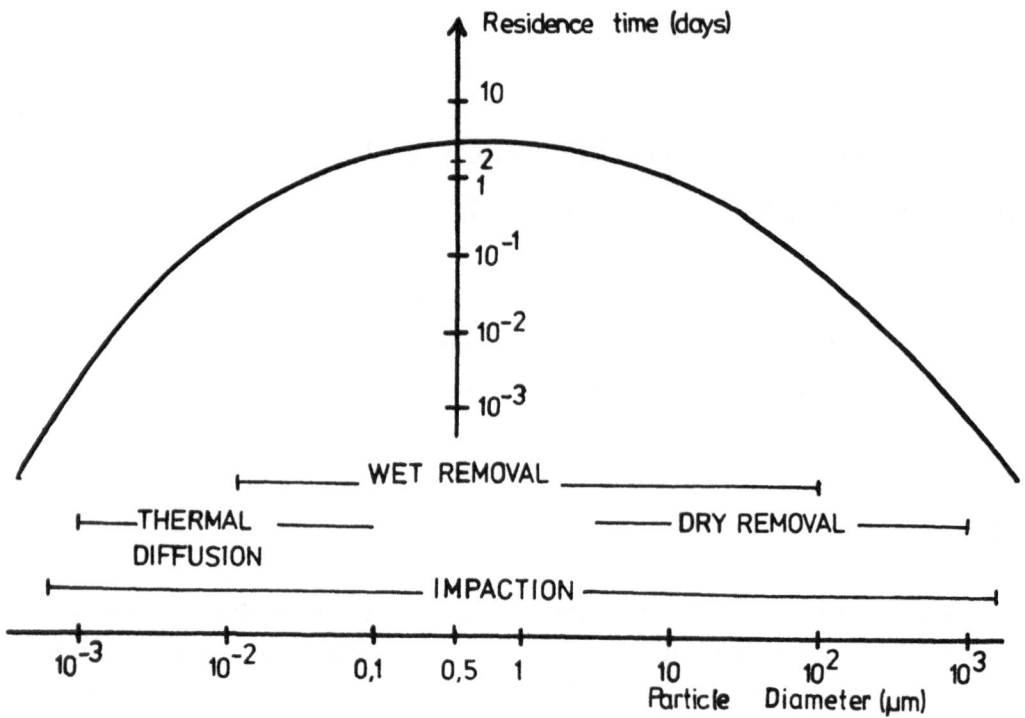

Fig.1: Residence time in the tropospheric mixing layer as a function
of particle size

The longer the residence time the farer the substance can be transported and the larger the atmosphere can be loaden with its mass.
For a toxic substance this are relevant aspects.

If we want to follow the pathway of chemical conversions of a substance the reactivity has to be taken into account. To each intermediate compound a residence time can be attached.

If we regard only elements or quasi-inert compounds the chemical
reactivity needs not to be considered. By this restriction the atmospheric cycles are better to handle.

2. EXPERIMENTAL BASIS

In order to determine the residence times of particle-bound substances impactor measurements have become a helpful tool. The
residence time of an airborne particle can be related to its size
(Jaenicke, 1978). This conception can be generalized to every particle-bound substance (Müller, 1981).

Impactor measurements in ambient air were carried out for total
particulate matter (Laskus, 1977) but also in respect to metals since

Fig.2:

Relative mass size dis-
tributions of metals and
total particulate matter
as a function of particle
size

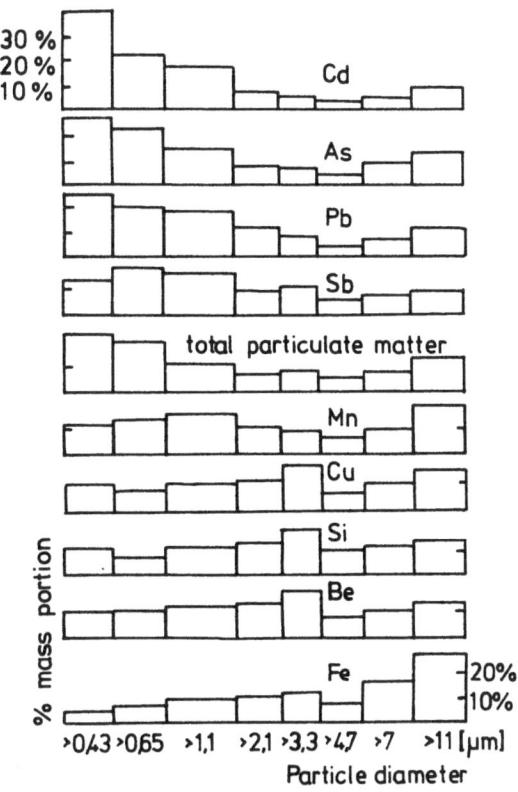

atomic absorption spectrometry is available (Müller, 1981; Wagner, Heidrun, 1981).

The important acid rain ions sulfate, nitrate and chloride which constitute about a quarter of total particulate matter are profoundly accessible in their mass size distributions since ion chromatography is available (Müller, 1981). Recent measurements yielded important results in the determination of the pH-value and the water-solubility of total particulate matter and several metals in respect to particle size (Wagner, W., 1982).

In conjunction with gas-phase and rain-phase measurements of sulfur- and nitrogen-compounds (Müller, 1981) an appreciable insight-look into the acid rain problem can be established. Mass size distributions of organic compounds up to now are limited to polycyclic aromatic compounds (PAH) (Van Vaeck and Van Cauwenberghe, 1981; Pflock, 1981). The distributions of the commonly volatile organics behave like comparable volatile inorganic compounds.

The cited measurements in Frankfurt mostly were carried out with an 8-stage Andersen-impactor. On the impactor stages glass discs for the sampling of the ambient air particles were used. During the sampling process the discs were turned several times in order to have loaden the whole collecting surface. Bounce effects from stage to

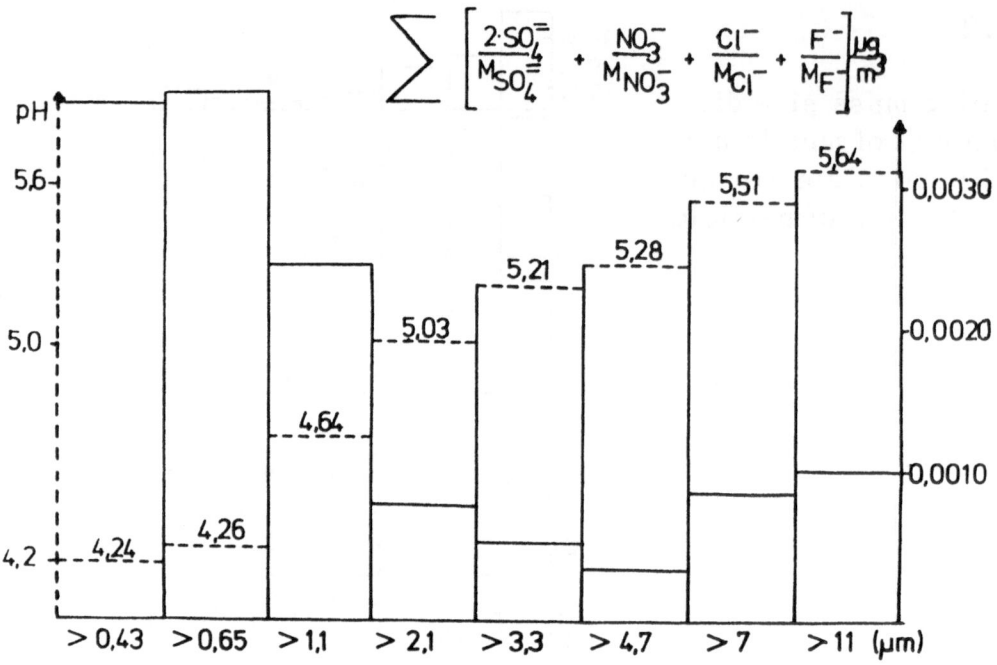

$$\sum \left[\frac{2 \cdot SO_4^=}{M_{SO_4^=}} + \frac{NO_3^-}{M_{NO_3^-}} + \frac{Cl^-}{M_{Cl^-}} + \frac{F^-}{M_F^-} \right] \frac{\mu g}{m^3}$$

Fig. 3: Mass size distributions of the sum of sulfate-, nitrate-, chloride- and fluoride-ions and of the pH as a function of particle size

stage cannot completely be avoided. The distributions of the substances relative to each other, however, are not influenced.

The particulate matter collected on the discs can be weighed and easily be rinsed off for analysis. Compared to loaden filters there is no additional material which complicates the analysis and diminishes the detection sensitivity.

3. THEORETICAL RELATION

The residence time of particles in the atmosphere is dependent on the removal processes. Nuclei mode (n.m.) particles ($<0,1\,\mu$m diameter, classification Whitby, 1975) behave similar to gases, dispose of high thermal diffusion and are quickly eliminated either by coagulation to bigger particles or deposited on the ground by impaction. Coarse mode (c.m.) particles ($>2\,\mu$m) are also quickly removed from the atmosphere by sedimentation in the course of hours.

In the range between 0,1 and 2 μm, the accumulation mode (acc.m.), the longest living particles exist with a maximum at 0,3 μm (Pruppacher, 1978). Impaction and wet processes ⇥ which are relevant for particles of all sizes - are the dominating removal processes for the

Fig. 4:

Water-solubility of iron,
total particulate matter and
cadmium as a function of
particle size

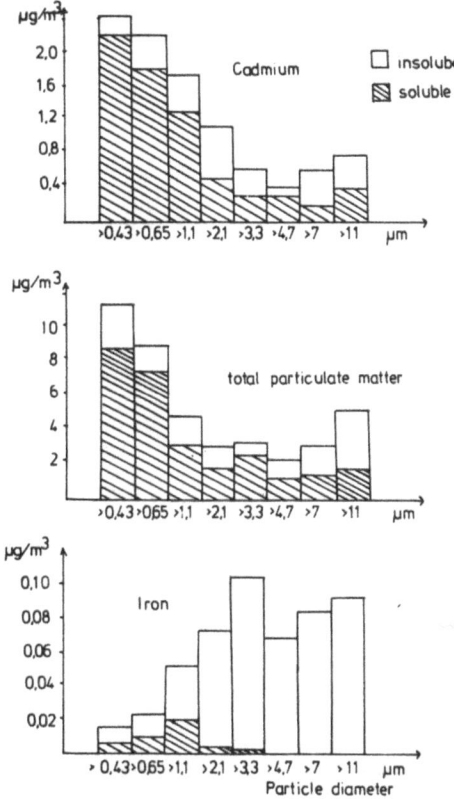

acc.m. particles. Thermal diffusion and sedimentation have only a
negligible influence on the removal of particles in this size range.

The residence time with exclusion of the impaction processes on the
ground can be formulated as a function of particle size (Jaenicke,
1978). The additional removal by impaction - which is also minimized
in the size range of acc.m. particles - has to be empirically supple-
mented. The deposition by impaction is dependent on the wind velocity
and the surface layer properties.

Nevertheless, by successive approaches a mean residence time curve
(Fig. 1) can be established and self-consistently be verified. For
every region such a residence time curve can be found and proved by
measurements.

In steady state of production E and removal I of the mass m of an
atmospheric substance, for the residence time τ exists the relation:

$$(1) \quad E = \frac{m}{\tau} = I$$

Regarding the flux F to the surface A of the earth, it can be written:

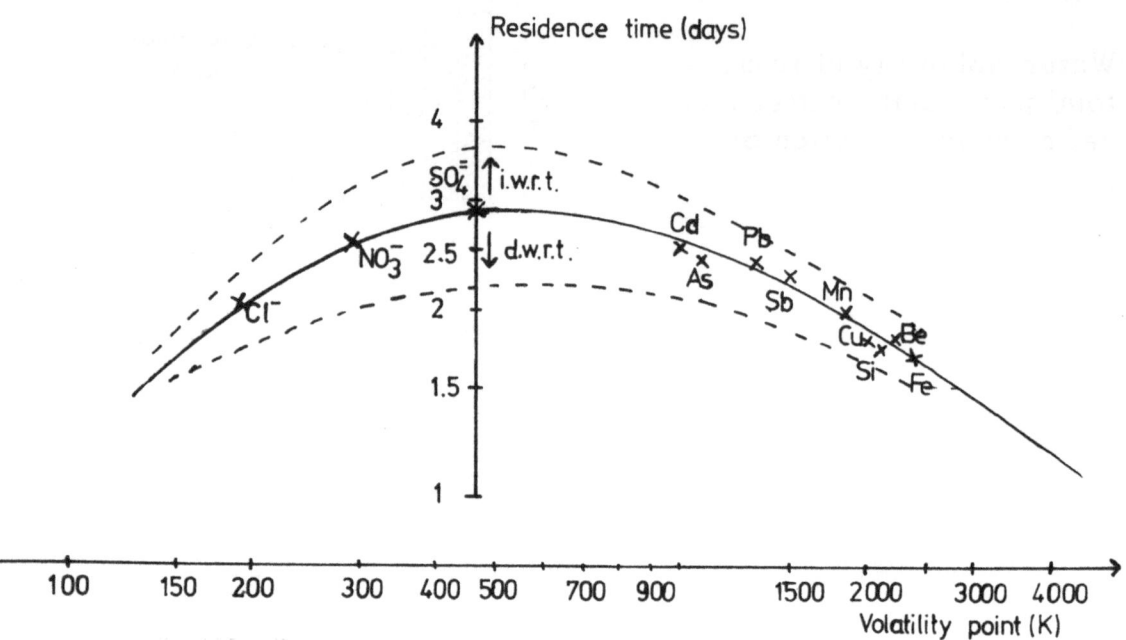

Fig. 5: Residence time of particle-bound substances in the tropospheric mixing layer as a function of the volatility point
(arithmetic mean of melting and boiling point)

$$(2) \quad F = \frac{I}{A} = \frac{m}{\tau \cdot A} = \frac{c \cdot H \cdot A}{\tau \cdot A} = c \cdot \frac{H}{\tau}$$

c is the mean atmospheric concentration of the substance within the
volume of the tropospheric mixing layer of height H between the
inversion and the ground level. The total deposition velocity v on the
ground is defined as v = F/c (Chamberlain, 1966). Compared with
equation (2), we get:

$$(3) \quad v = \frac{H}{\tau}$$

The deposition velocity is inversely proportional to the residence
time. A long residence time of an atmospheric substance corresponds
to a small deposition velocity.

The residence time τ of a particle-bound substance can be determined
from its measured mass size distribution by use of the curve in Fig. 1.
To each size fraction a mean residence time τ_i can be attached and
by summation over all mass fractions X_i, we get:

$$(4) \quad \tau = \sum X_i \cdot \tau_i ; \qquad \sum X_i = 1$$

The mass size distribution of a substance is largely influenced by its vapour pressure. The more volatile a substance is the more mass accumulates in smaller particles. More mass in smaller particles induces a larger surface for exchange with the ambient gas phase. Volatile substances are enriched at the surface of particles which originate from combustion processes (Biermann and Ondov, 1980).

However, the vapour pressures of most solid substances under atmospheric conditions are very low (for heavy metals in the order of 10 high -10 to -15 Torr) and only a few data are available. Therefore the boiling and melting point are used as auxiliary parameters. The arithmetic mean value, the volatility point T_v can be related to the residence time (Müller, 1981):

$$(5) \quad \ln \tau = -\tau_1 T_v + \tau_0$$

The equation is applicable for substances which have their mass in the acc.m. and the c.m. of airborne particles. The constants τ_1 and τ_0 are essentially influenced by the wet residence time. The values of metals determined with equation (4) fit into equation (5).

The short living n.m. and c.m. particles are mostly removed by dry deposition, the acc.m. particles predominantly by wet-processes. By knowledge of the mean mass size distribution of a substance its wet to dry deposited ratio I_{wet}/I_{dry} can be estimated:

$$(6) \quad \frac{I_{wet}}{I_{dry}} = \frac{X_{acc}}{X_m + X_c}$$

$$X_m + X_{acc} + X_c = 1$$

X_m, X_{acc}, X_c = mass fractions in the n.m., acc.m. and c.m.

4. DISCUSSION OF MEASUREMENTS

4.1 Metals

In Fig. 2 the measured mass size distributions of metals are represented. Instead of the concentrations the mass percentages of the stages are plotted versus particle size. The total collected mass of each substance is set 100%. The plot of the relative mass distributions make the metals better comparable. By use of concentrations there are differences in orders of magnitude. It was observed that in spite of the largely varying concentrations during the seasons the relative distribution of a substance stays reasonably stable.

The corresponding residence times determined with equation (4) are plotted in Fig. 5. The volatile metals like Cd or Pb have most of their mass in the acc.mode. Therefore they stay longer in the atmosphere and large amounts of the volatile metals are transported far away from their sources. According to equation (6) these substances are enriched in wet deposition, in which 4/5 of the deposited mass is found (Rohbock et al, 1981).

Heavy volatile elements like Si or Fe have their mass predominantly in the c.mode. Therefore, they have shorter residence times due to sedimentation. The elements mainly situated in c.m. particles are enriched in dry-deposition.

4.2 Strong acid ions and solubility

By aid of ion-chromatography the sulfate-, nitrate-, chloride- and fluoride-ions as well as the pH were measured as a function of particle size (Wagner, W., 1982).

The highest acidity was found in the acc.mode (Fig. 3). The c.m.particles have a pH of 1-2 scales higher. In n.m.particles the pH also increases compared to the acc.mode as was observed by analysis of back-up filters of the impactor. The pH-distribution correlates with the sum of the four ions normalized with their molecular weights. $SO_4^=$, NO_3^- and Cl^- are responsible for about 90% of the acidity. In the acc.m. $SO_4^=$ is reliable for about 3/4 of the acid substances whereas in the other modes the sum of NO_3^- and Cl^- contributes comparably high. The acc.m. particles which are mostly removed by wet processes transform their acidity into the rainwater.

The pH of an airborne particle is a function of its size. The solubility of a particle-bound substance, therefore, is largely influenced by its mass size distribution. The more mass is situated in the acc.m. the higher are the water-soluble portions. Sulfate which is almost entirely bound in this mode is soluble by more than 95%. In Fig. 4 the water-soluble portions of total particulate matter, cadmium and iron are represented (Wagner,W,1982). Cd bound in the acc.mode is almost entirely soluble. Fe bound mainly in the c.m. has only little soluble quantities which are situated in the acc.m. where the pH is sufficiently low.

Consequently, according to the removal processes the solubility of a substance in the different atmospheric phases is changing. Compared to airborne particulate matter the solubility is lower in the dry deposited and higher in the wet deposited matter (Rohbock et al., 1981). The soluble substances are enriched in wet and the less soluble ones in dry deposition.

4.3 Classification of substances

In Fig. 5 the residence times determined with equation (4) are plotted versus the volatility points. By use of logarithmic scales a fairly symetric plot for particle-bound substances is obtained.

Sulfate has vapour pressure properties which enable it to be the particle-bound substance with the maximum of residence time. Sulfate, therefore, is the typical long-range pollutant. Sulfuric acid preferentially settles down in the center of the acc.m. and pushes away the more volatile nitric and hydrochloric acids.

To a smaller extent the volatile metals As, Cd, Pb and Sb and NO_3^- can also be regarded as long range pollutants. Within a distance of about 1000 km their deposition rates correlate with sulfate. Under Middle-European conditions, for total sulfur a residence time of 2,2 and for NO_x-nitrogen of 1,6 days was estimated (Müller, 1981).

Metals like Mn and Cu are spread over all modes and can be classified as regional pollutants. Similar properties can be attributed to chloride which has about 1/3 of its mass in each of the three modes. In contrast to $SO_4^=$ and NO_3^-, therefore, large quantities of Cl^- are dry deposited.

Heavy volatile minerals, like Si and Fe, bound in c.m.particles have short residence times. They can be regarded as local or regional pollutants.

The absolute height of the residence time curve in Fig. 5 is largely dependent on the wet residence time. In global latitudes with many rainfalls the absolute values decrease. The curve is flattened and the differences between the substances become less significant. The wet removal dominates in the atmospheric scavenging processes.

In an aride climate (increased wet residence time) the curve is lifted to higher absolute values. As a consequence the differences of residence time between the substances are significantly influenced by their volatility properties.

REF.:

Biermann, A. H. and Ondov, J. M.: 1980, Atmospheric Environment 14, pp. 289-295
Chamberlain, A. C.: 1966, Proc. Roy. Soc. A 290, pp. 236-265
Jaenicke, R.: 1978, Ber. Bunsenges. Phys. Chemie 82, pp.1198-1202
Laskus, L.: 1977, Staub-Reinhalt. Luft 37, pp. 299-306
Müller, J.: 1981, Proc. "Heavy Metals in the Env.", Amsterdam, c/o CEP Consultants Ltd, Edinburgh EH1 3QH, UK
Müller, J.: 1981, Proc. "Second European Symposium on Physico-Chemical Behaviour of Atmospheric Pollutants", Varese (It.), c/o CEC, Brussels
Müller, J.: 1981, Proc. "VDI-Kolloquium Schwebstoffe und Stäube", Nürnberg, c/o VDI-Düsseldorf
Pflock, H.: 1981, Diplomarbeit, c/o Inst. Meteorologie und Geophysik, Universität Frankfurt (M.)
Rohbock, E., Georgii, H.-W., Perseke, C., Kins, L.: 1981, Proc. "Heavy Metals in the Env., Amsterdam, c/o CEP Consultants Ltd, Edinburgh EH1 3QH, UK
Rohbock, E., Georgii, H.-W., Perseke, C.: 1981, Proc. "9. GAF-Conference, Duisburg, c/o W. Stöber, Schmallenberg/Germ. pp. 80-85
Van Vaeck, L. and Van Cauwenberghe: 1981, Proc. "Second European Symposium on Physico-Chemical Behaviour of Atmospheric Pollutants", Varese (It.), c/o CEC, Brussels
Wagner, H.: 1981, Diplomarbeit, c/o Inst. Meteorologie und Geophysik, Universität Frankfurt (M.)
Wagner, W.: 1982, Diplomarbeit, c/o Inst. Meteorologie und Geophysik, Universität Frankfurt (M.)
Whitby, K. T.: 1975, Progress Report EPA R 800971

Wet Deposition

GLOBAL DISTRIBUTION OF THE ACIDITY IN PRECIPITATION

Hans-Walter Georgii
Department of Meteorology and Geophysics
University of Frankfurt/Main, FRG

ABSTRACT: The spatial distribution of pH-values in precipitation has been evaluated on the basis of measurements
at stations of the precipitation chemistry network of the
World Meteorological Organisation for the years 1972 to
1979. A conclusive picture of the global distribution cannot be given but the contours of precipitation-acidity gain
shape for Europe and North America.

INTRODUCTION

Upon the initiative of WMO and UNEP (United Nations Environment Program) a precipitation-chemistry network was
established which began operation in 1972. From that year
onwards an increasing number of countries and stations par-
ticipated in this undertaking. Now, approximately 110
stations report the results of the analysis of several compounds like pH, conductivity, strong acids, potassium, magnesium, calcium, chloride, ammonium, nitrate, sulfate and
certain heavy metals to the "WMO Collaborating Center on
Background Air Pollution Data" which is located at the Environmental Protection Agency (EPA) in the USA. Data were
published for the years 1972 to 1978 and on a limited basis
for 1979. With respect to the integrity of the data on
chemical compounds in rainwater, three series of intercom-
parisons using samples of known composition which were
distributed to 32 participating laboratories, were orga-
nised. The results show a great improvement of the performance of the laboratories from the first to the third test-
series, the variability of the data furnished by the individual laboratories decreased. However, for sulfate and
nitrate the variation is still considerable. The evaluations of the test-series give only an indication with respect to the performance of the laboratories as a group.
It is, however, not possible to draw any conclusions with

H.-W. Georgii and J. Pankrath (eds.), Deposition of Atmospheric Pollutants, 55–66.
Copyright © 1982 by D. Reidel Publishing Company.

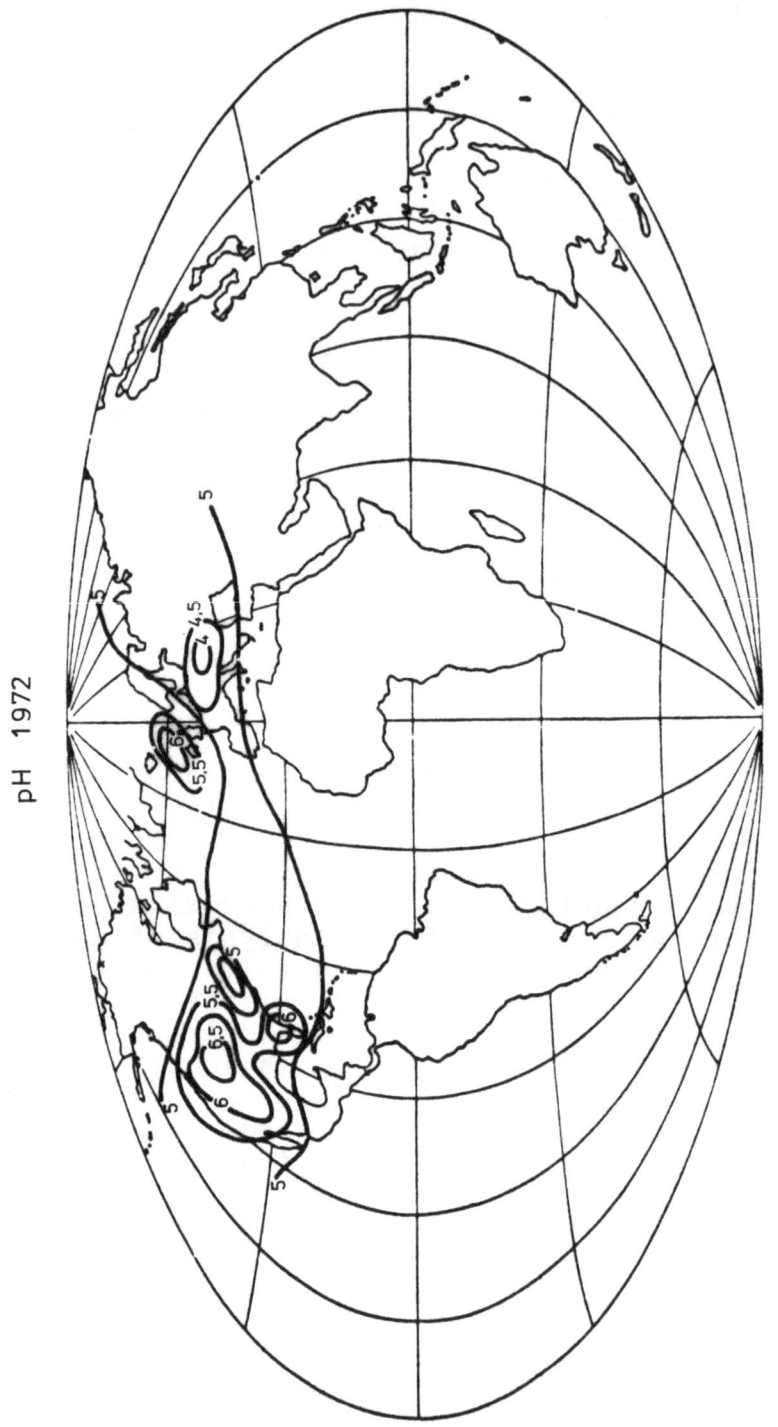

pH 1972

Fig.: 1: Spatial distribution of pH in rain for 1972

regard to the continued analytical performance of any
individual laboratory. A. Köhler (1980) mentions that
the laboratory-error is certainly not the only existing
one. Probably, large variations exist in sampling procedure and in siting errors produced by insufficient repre-
sentativeness of sampling sites.

SPATIAL DISTRIBUTION OF PH IN PRECIPITATION

In a recent study H.-W. Georgii (1981) has reviewed the
distribution of acidity in rain on the basis of WMO-data
for the years 1972 to 1978. This evaluation can now be
updated for 1979. BAP Mon-data should permit to draw a
global picture of the distribution of different chemical
compounds. In context with the problem of the so-called
"acid rain", the pH-value is one parameter of special interest. Fig. 1 shows the spatial distribution of pH for the
year 1972. It can be seen that in Central Europe pH-values
as low as 4 are found with increasing values towards the
oceans but also towards Eastern Europe. Over the North
American continent relatively high pH-values were observed
over the middle and Western part of the US. Only at the
East coast the acidity of the rain leads to pH of 4.5.
From 1972 to 1979 the number of reporting stations has considerably increased. It should be mentioned that from comparison between the distribution of pH-values in 1976 and
1978 it was found that the acidity in the United States
showed some increase both in the Central and Eastern states
while there was no significant change in Europe. Fig. 2
shows the pH-distribution for the year 1979. There is no
striking change in Europe compared with the previous years;
in the Eastern part of the US the pH-values are even somewhat higher in 1979. During that year it was possible to
include precipitation-chemistry data for India, Pakistan
and other Asian countries which show surprisingly high pH-
values. Assuming the high pH-values are correct - there
were indications for high pH already in 1978 - they can
only be explained by the dispersion of alkaline aerosols
from the ground which are transported in the atmosphere
and washed out by precipitation. From the study of the
pH-distribution over North America and Europe for the
period 1972 to 1979 it appears that obviously the acidity
in rain did not generally increase. However, it is still
not possible to draw any conclusions regarding the global
picture due to lack of stations in other continents than
North America and Europe. Stations in the Southern hemisphere are still missing. It could also be appropriate to
install measuring sites on islands in the Atlantic and
Pacific oceans to permit a better interpretation of inter-
hemispheric transport processes.

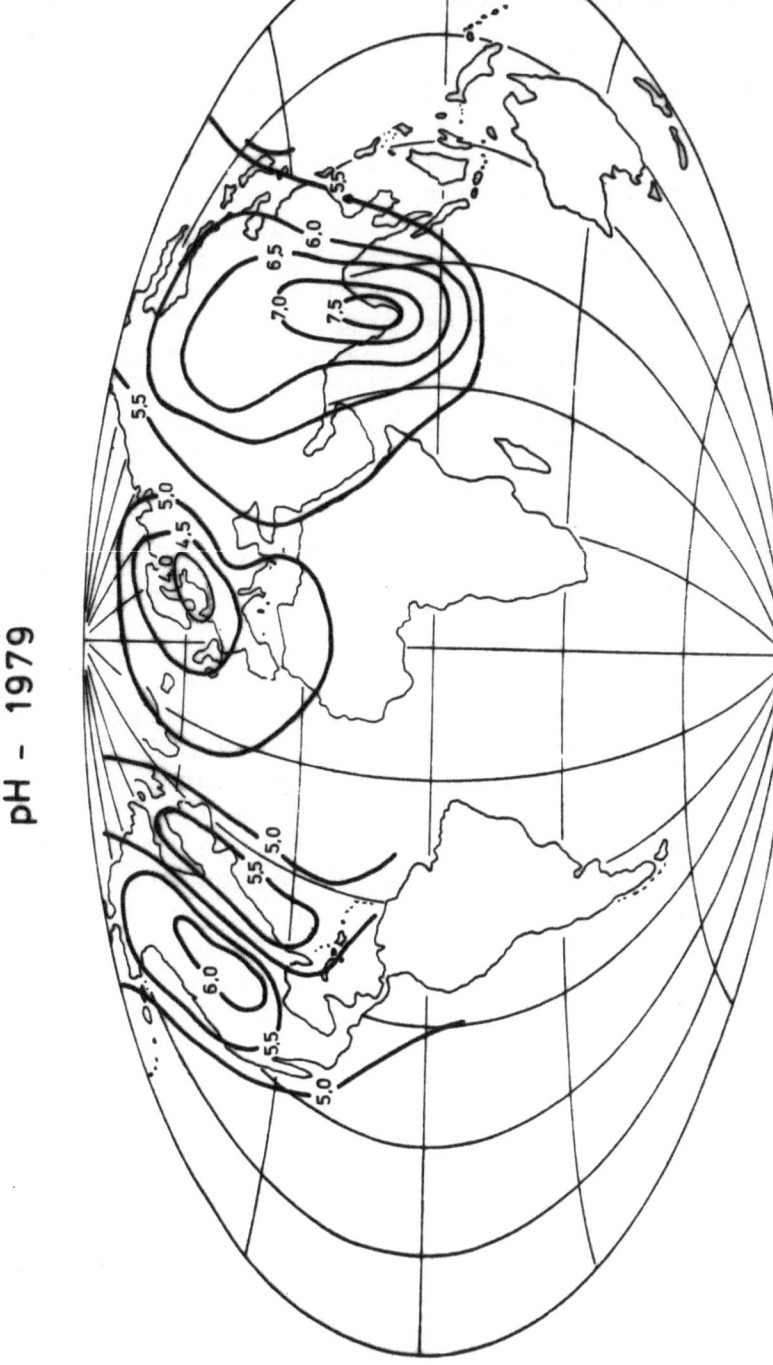

pH – 1979

Fig.2:   Spatial distribution of pH in rain for 1979

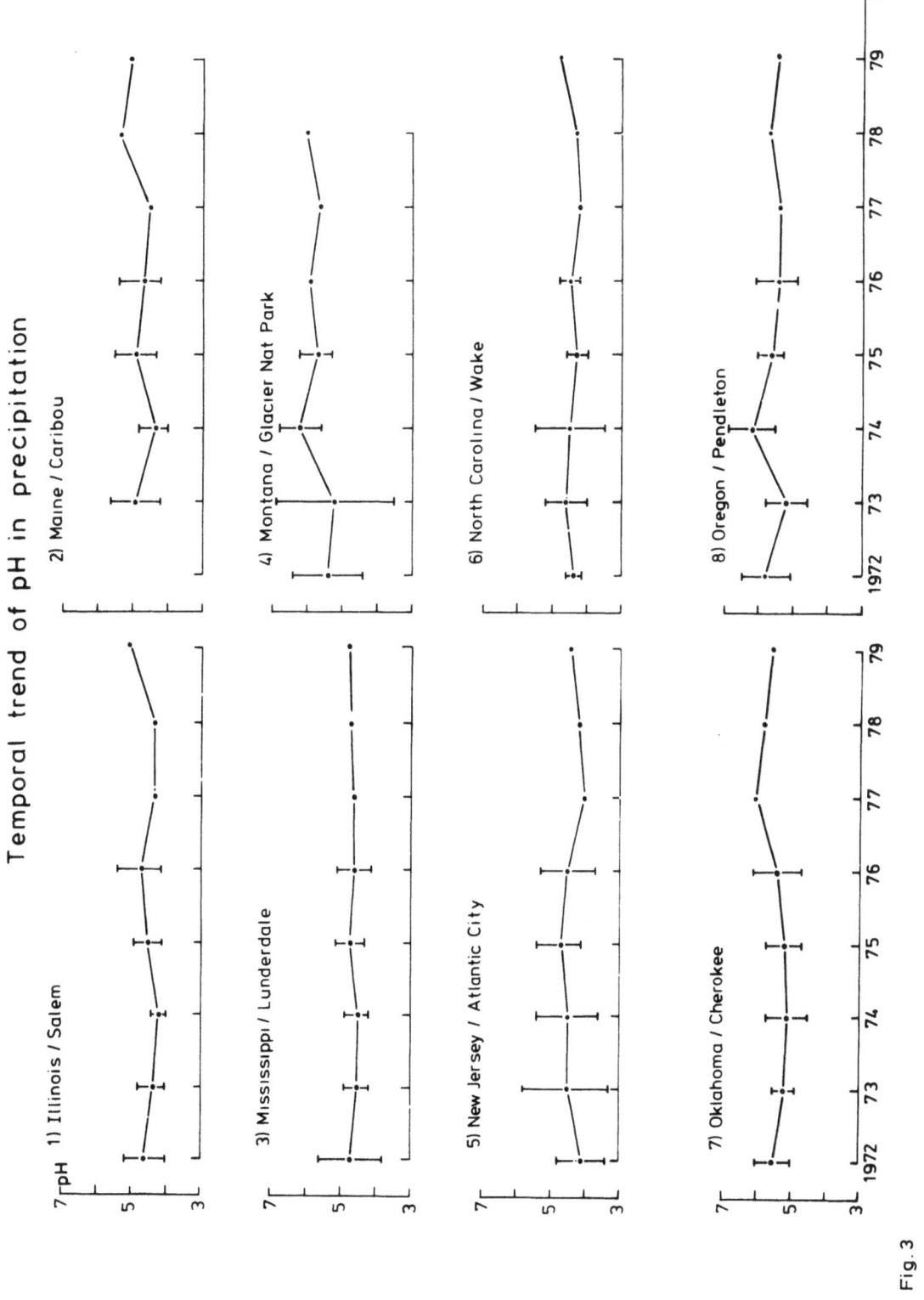

Fig. 3

TEMPORAL TREND OF ACIDITY IN RAIN-WATER

Kayser et al. (1974) found, when evaluating the composition of precipitation at the German mountain-station Schauinsland a decrease of the pH-value in rain from 5.0 to
about 4.2 during the period 1965-1973. The mean pH-value
averaged over 8 stations in West Germany analysed by the
same authors for the period 1967 to 1972 dropped from 4.7
to 4.3. An evaluation of 23 stations of the BAP Mon-pre-
cipitation chemistry network for the period 1972-1979 per-
mits some conclusions with respect to the trend of the pH-
value in precipitation during that eight-year period.
Fig. 3, showing the temporal trend for these stations, suggests an increasing pH-value for some of the recent years,
while Tom Green in Texas shows a trend towards higher acidity. The West German station Langenbrügge shows a tendency towards higher acidity while on Mt. Schauinsland and on
Mt. Brotjacklriegl pH-values suggest a decrease of the
rain-acidity. It is interesting to note that the two Swedish stations do not reveal a steady drop of pH in rain
during the period 1972-1979. The data do not support a
generalisation of decreasing pH-values in rain in North
America and in Western Europe. It appears that the tendency of increasing resp. decreasing acidity and its variation
from year to year is more a regional problem depending largely on the changing production, transport and scavenging
of pollutants but also on the year-to-year variation of
the rainfall-rate. This statement is supported by an evaluation by Miller published in a paper by Pack (1979).
Miller found no trend of the pH in precipitation at five
WMO-regional stations in the USA over the period 1972-1976.
However, one should be careful in making generalizing
statements since an eight-year period is too short for conclusive observations.
Besides the temporal trend of the pH-value we have also
evaluated the frequency distribution of pH in precipitation
for eight stations following a publication by Kasina (1979).
Figs. 4 and 5 show the results for six stations in Europe
and in North America. In this figures we have compared
the period 1972-1976 (solid lines) with the period 1977-
1980 (dotted lines). It can be seen that at some of the
stations the 50%-value of the frequency distribution has
moved to lower pH-values during the years 1977-1980. In
contrast to that at some stations the 50%-value has increased. The picture is not very conclusive, it can only
be stated that the trend of pH shows distinct regional differences.

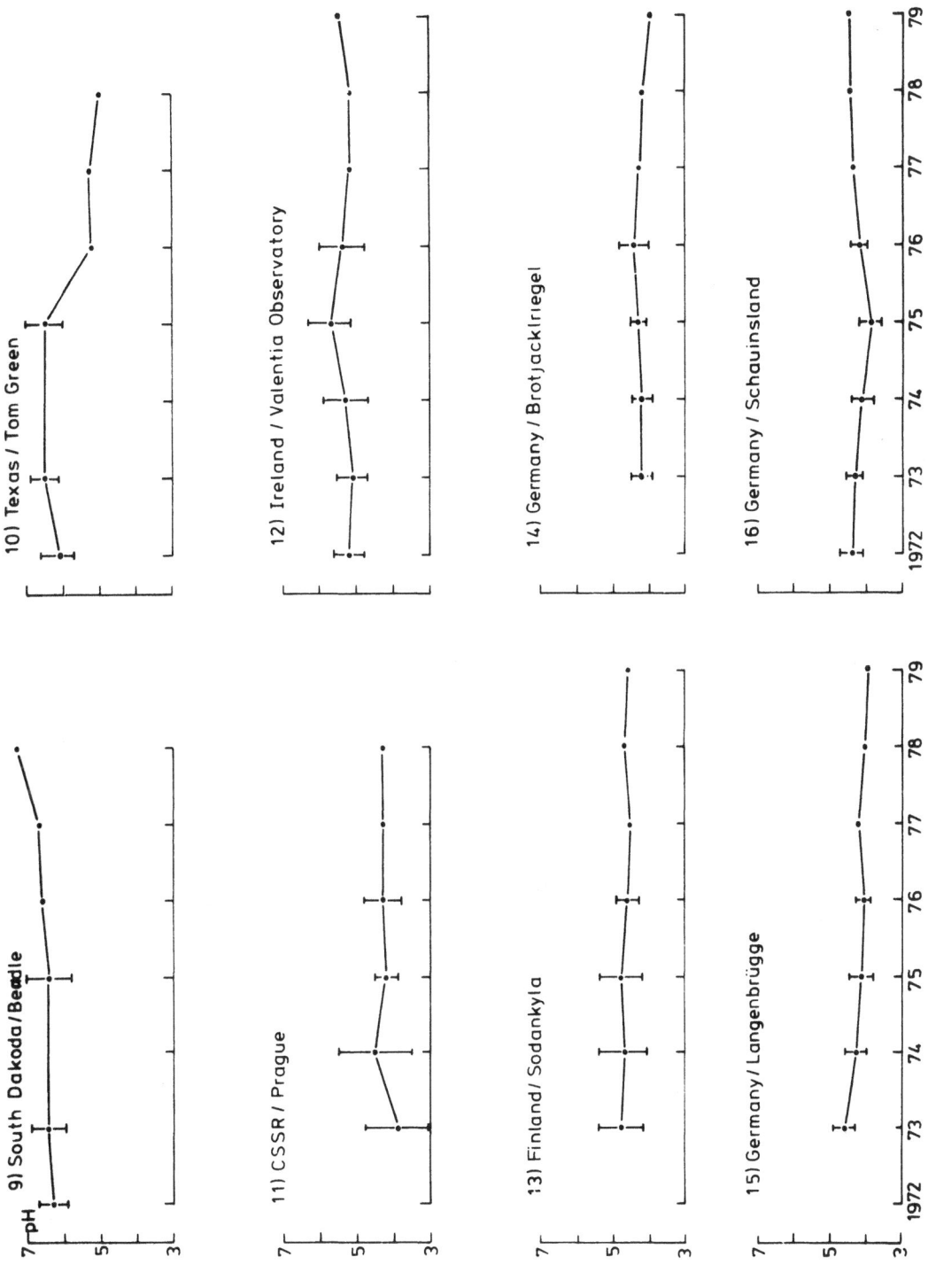

Fig. 3

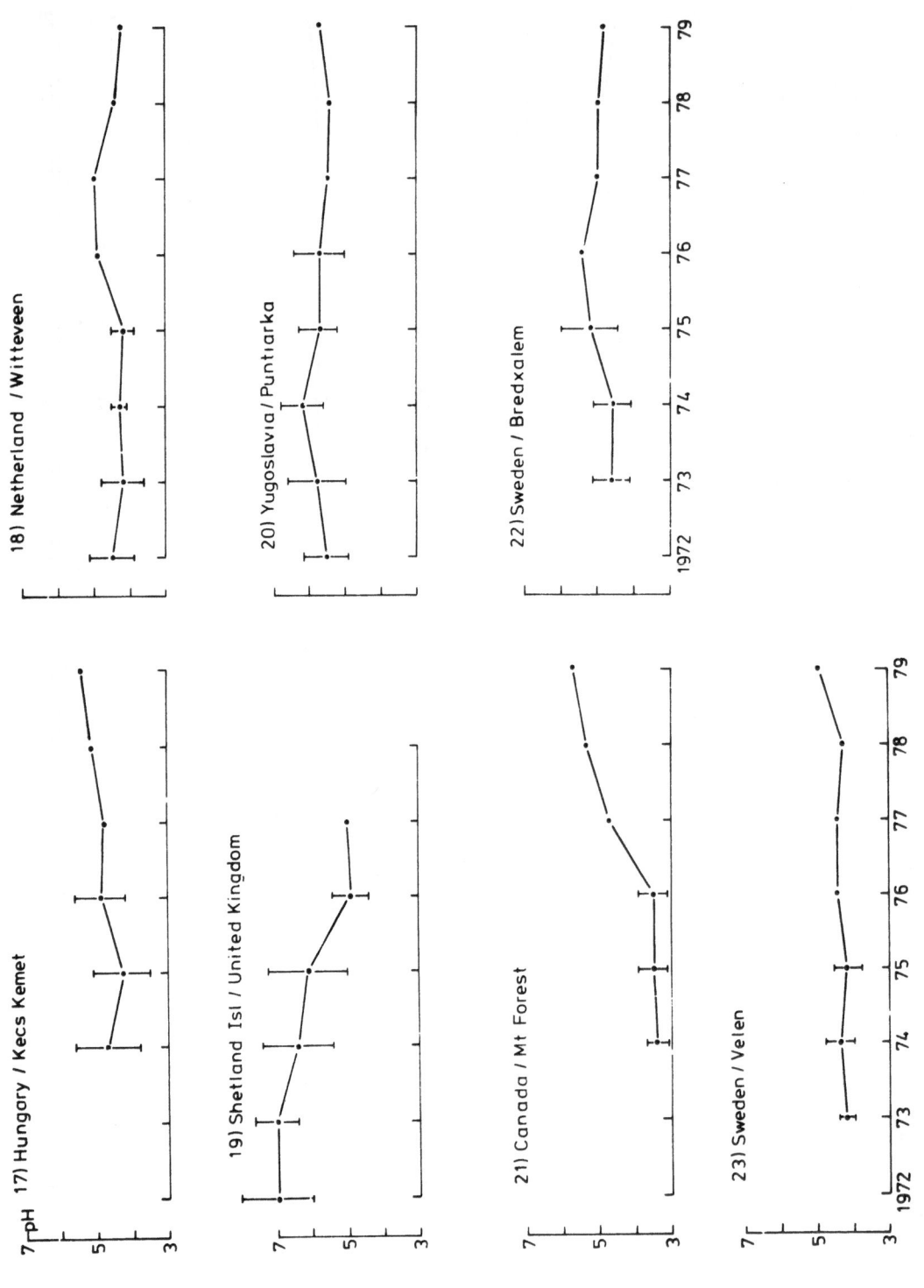

Fig. 3

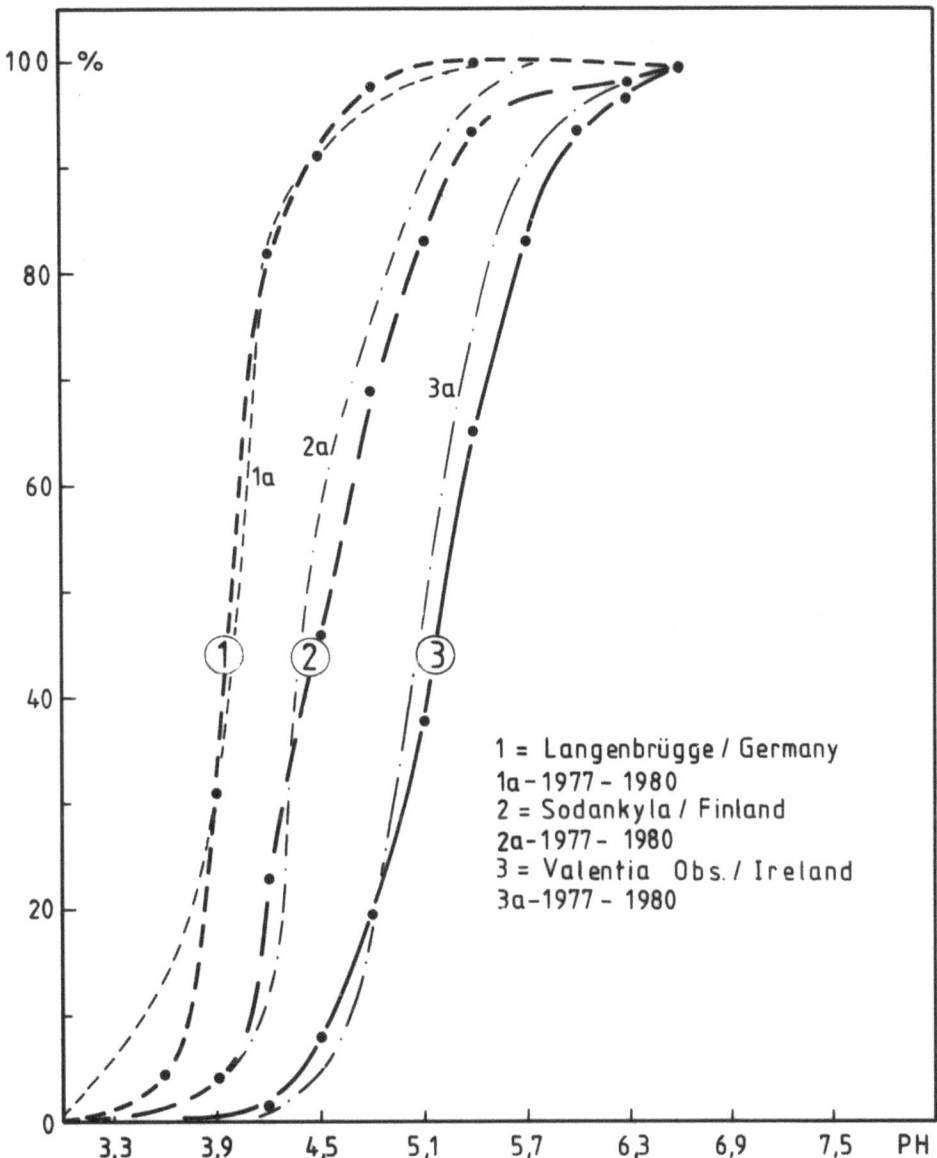

1 = Langenbrügge / Germany
1a - 1977 - 1980
2 = Sodankyla / Finland
2a - 1977 - 1980
3 = Valentia Obs./ Ireland
3a - 1977 - 1980

Frequency - distribution of PH in precipitation

Fig. 4

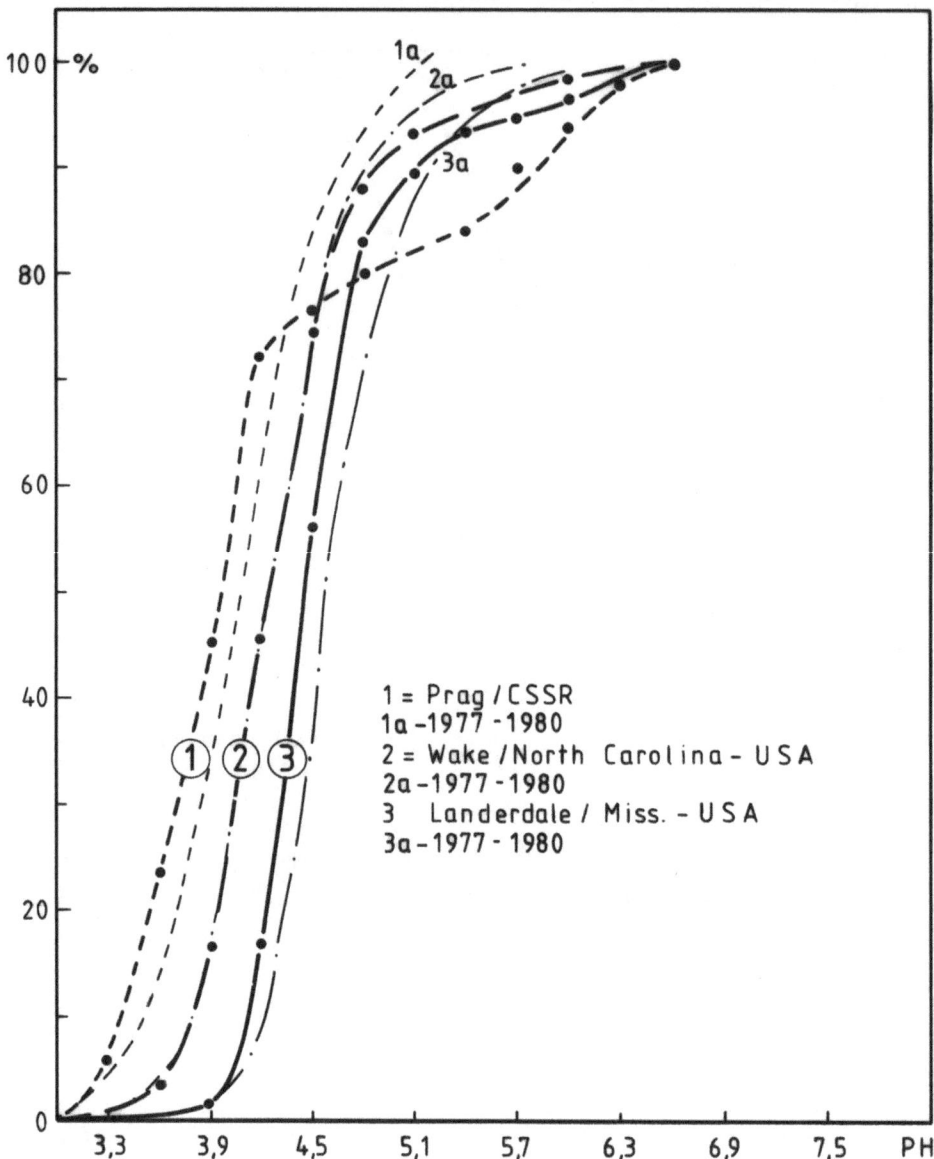

Frequency distribution of PH in precipitation

Fig. 5

SULFATE DISTRIBUTION IN PRECIPITATION

The $SO_4^=$ distribution in rain has gained growing attention during the last years, since sulfate is supposed to be one of the dominant components in the production of "acid rain". From the WMO-data base it appears that the sulfate-concentration over Europe and North America did not change significantly but a further increase in sulfate was noticed from 1976 to 1978 both in Europe and the USA. Finally, there are indications that the amount of sulfate in precipitation over the North Atlantic ocean has sharply increased since 1972. Although data do not allow a definite conclusion, it is hinted that long range transport of sulfate from the North American continent may be of importance over the Atlantic but also influence the sulfate-concentration over Western Europe.
C. C. Wallen (1980) has calculated mean values of the sulfate-concentration in rain for the period 1972-1976 and compared them with values for the late 1950's for Europe which were published by deBary and Junge (1963). Wallen comes to the following conclusions:

1) There was hardly any increase in sulfate content in rain in northern Scandinavia from the fifties to the seventies
2) In southern Scandinavia the increase is of the order 50 to 100%, that means from 1.0 to 1.5 mg/l
3) In the maximum area of Europe the increase is also about 100%, that means from 2.3 to 4.5 mg/l
4) In south and south-eastern Europe the mean values of sulfate increased from the fifties to the seventies by about 50% from 1.3 to 1.9 mg/l.

Comparing the average sulfate-concentration for Europe based on the data by deBary and Junge with the evaluations by Wallen, there is an indication for an average increase for the whole area of about 50%.

CONCLUSIONS

The quality of the data of the WMO-network may be questionable in some cases and the operation period since 1972 is too short yet to allow for assessments of definite trends, the continuation and expansion of the network seems essential to obtain a global overview of the situation of chemical composition of precipitation. In spite of the existence of the global network and as a necessary addition, regional networks are required for the study of regional trends, for the study of the transport of pollutants and for the study of deposition patterns. Such networks have

been installed and are operated in an increasing number
of countries in North America and in Europe.

REFERENCES

de Bary, E., and Junge, C. (1963) Distribution of Sulfur
                 and Chlorine over Europe
                 Tellus 15, 370

Georgii, H.-W. (1981) Review of the acidity of precipi-
                 tation according to the WMO-network
                 Idöjaras 85, 1-10

Kasina, S.(1979) Precipitation Acidity in the Krakow-
                 region
                 EPRI-Report SOA 77-403

Kayser, K., Jessel, U., Köhler, A., Rönicke, G. (1974)
                 Die pH-Werte des Niederschlages in der
                 BR Deutschland 1967-1972
                 Kommission z. Erforschung der Luftver-
                 unreinigung, Mitteilung IX

Köhler, A. (1980) The WMO BAPMon Proc. Symposium on the
                 Development of Multi-Media Monitoring of
                 Environmental Pollution, Riga 1978
                 WMO-Report No. 563, 37-49

Pack, D.H. (1979) Acid Precipitation - The Physical System
                 Proc. Advisory Workshop on the formation
                 of Acid Precipitation
                 EPRI Report No. EA 1074, 3-37

Wallen, C. C. (1980) A preliminary Evaluation of the
                 WMO-UNEP Precipitation Chemistry Data
                 MARC-Report No. 22

DEPOSITION OF ACID IN PRECIPITATION

Peter Winkler
Deutscher Wetterdienst, Meteorologisches Observato-
rium Hamburg Frahmredder 95, D-2000 Hamburg 65

ABSTRACT

From the pH and electrical conductivity measured with the
automatic precipitation analyser the fraction of total
strong acids in precipitation water is derived. Below a pH
value of 4,3 the dissolved matter is dominated by acid. At
higher pH values the acid fraction decreases rapidly. Com-
parative studies of the acidity content of aerosol particles
showed that the acid of aerosol is by far not sufficient to
explain precipitation acidity. - For selected events the
influence of the variation of rain drop size spectra on the
trace substance content of precipitation is demonstrated.
- Average depositions for different air masses as derived
from trajectory analysis are presented. The most astonishing
result is that for transport over the North Sea without
land contact the deposition is nearly as high as for transport over the highly industrialized Ruhr area. - The long
time trend of precipitation pH is discusses.

1. INTRODUCTION

For biologists, soil scientists and ecologists the deposition of substances is of higher interest than the concentration of these substances in the air or in the precipitation. Because precipitation is a relatively seldom event
as compared with the time free of precipitation dry deposition can for many substances give an important contribution to the total deposition. In the case of acid it can
be shown that precipitation brings the highest contributions so that we can restrict to the relative composition
of the precipitation. Deposition is then simpoly obtained
by multiplying the concentrations with the precipitation
amount.

H.-W. Georgii and J. Pankrath (eds.), Deposition of Atmospheric Pollutants, 67–76.

2. METHOD OF MEASUREMENT

For measuring the pH and the electrical conductivity of the
precipitation an automatic precipitation analyser has been
developed (Fig.1) (Winkler,1977). The polyethylen funnel
is covered by a lid opening during precipitation events
only. From the funnel the rain water flows through a conductivity cell to the pH probe. From here it leaves the
instrument over a tipping bucket which measures precipitation amount and intensity.

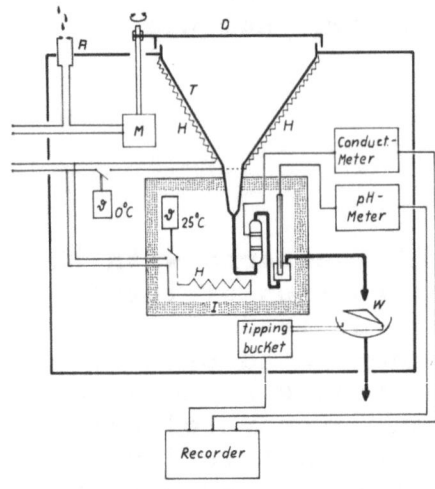

Fig.1 Automatic precipitation
analyser. R = rain sensor,
D = motor driven lid, M = Motor,
W = tipping bucket, T = polyethylen funnel H = heating,
θ = thermostat I = isolation

The record is evaluated on a hourly bases. Values are taken
as precipitation weighed concentrations, which in the case
of pH are converted back to the average pH.

The electrical conductivity corrected for the influence of
the H^+ ions is a measure for the total amount of soluble
inorganic ions. By assuming an average equivalent conductivity (60 uS/cm) and am average equivalent weight (35) we
can assess the mass of dissolved matter. From the pH the
total amount of strong acids is estimated. In a pH - conductivity diagram (compare Fig.2) a pure diluted acid
droplet produces a certain pH and a distinct electrical
conductivity. At a fixed pH the conductivity must have a
minimum value produced by the H^+ ions alone. Lower conductivity values are forbidden. Additional dissolved ions
increase the conductivity letting the pH unchanged. So a
set of inclined lines is calculated which describe the
fraction of free acid contributing to the total dissolved
material. In such a diagram any dilution or concentration
of a solution due to evaporation or condensation means a
movement parallel to the inclined lines.

3. GENERAL RESULTS

Average results for Hamburg precipitation are depicted in
fig.2. To calculate the average values, we have averaged
the conductivities weighed with the precipitation amounts.
This method corresponds to a set of different collection
bottles each labeled with a distinct pH value. Each bottle
openes and collects precipitation only when the pH has the
right value. The average conductivity calculated corresponds
now to the conductivity of such a bottle which is measured
after a long time.

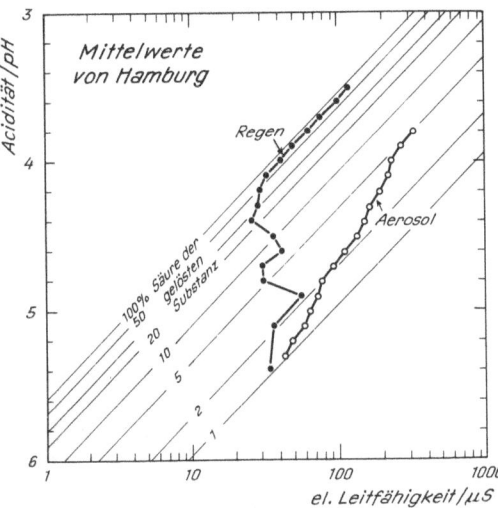

Fig 2 Average pH and electrical
conductivities of the precipitation of Hamburg (upper curve)
and of aerosol leaching solutions
(particles r > 0.1 μm, leached
with 0.4 ml H2O per m³ air
collected) (lower curve). The
inclined lines give the percentage
of free acid contributing to the
total dissolved material.

We see that below a pH value of about 4.3 the dissolved
material is dominated by free acid. At higher pH values
the acid fraction lowers rapidly.

For reasons of comparison we show in fig.2 the average acid
content of aerosol particles (r > 0.1 μm). These particles
were collected with an impactor. The samples were leached
with several ml of water so that the aerosol of one m³ of
air was dissolved in 0.4 ml H2O. This amount is about
comparable with the liquid water content of clouds and
therefore the leaching solutions are comparable with rain
water. It can be seen that on the average the acid fraction
of the aerosol is around 2% which does contribute nearly
nothing to precipitation acidity. A high washout or rainout
of aerosolparticles would indeed decrease the pH value of
the precipitation but it increases also the conductivity
so that those processes cannot increase the relative amount
of acid contributing to the dissolved material.

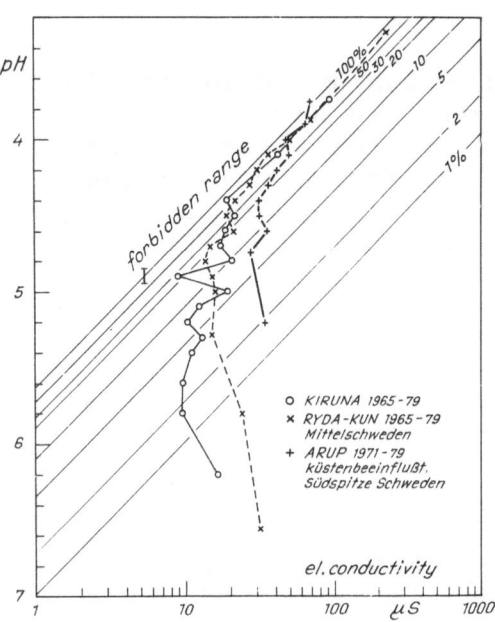

Fig 3 Averaged pH values and electrical conductivities of three swedish stations (Granat 1980). Compare the similarity of the characteristic lines of these stations with those of Hamburg.

It is most interesting that the precipitation of other stations shows a very similar characteristic relation between pH and electrical conductivity. In fig.3 the characteristics of three swedish stations are depicted (Granat 1980). We see that for the stations Kiruna (northern Sweden) and Ryda-Kun (central Sweden) for pH values below 4.6 the precipitation is dominated by acid and that at higher pH values the acid fraction lowers rapidly. The station Arup behaves different due to coastal influence.

The leg of the characteristic where the acid fraction lowers rapidly is related to a smaller conductivity for Kiruna than for Ryda-Kun and this in turn for Hamburg. This may reflect the influence of washout which can be expected to be higher in Hamburg smaller in Ryda-Kun and lowest in Kiruna.

The special importance of the characteristic pH - conductivity relation and its similarity for different stations is to be seen in the light that it averages atmospheric conditions where the amount of trace substances offered to the precipitation is in equilibrium with all processes as wash out, rain out, vertical gradients, microphysics, chemical processes and so on which all influence the removal of trace substances by precipitation. In the case of acidity there exists obviously such a characteristic equilibrium relation. It is still an open question what are the reasons for the two different legs.

Additional information can be obtained by trajectory analysis. The 48 hours trajectories (850 mbar) were clas-

sified into 8 sectors. For each sector the corresponding
pH and conductivity values were averaged by weighing them
with the precipitation amounts. The result is presented
in fig 4.

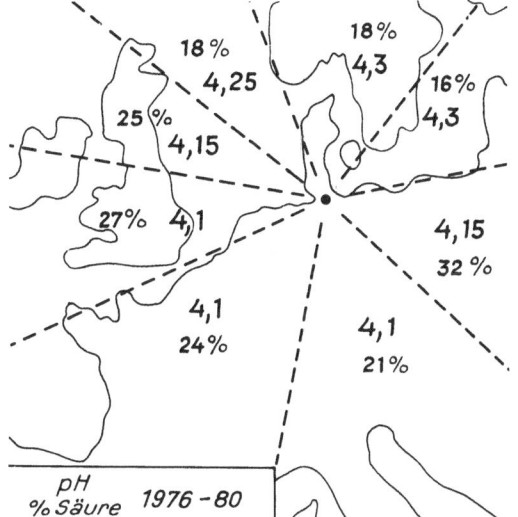

Fig 4 Average pH values conductivities and percentage of free
acid as related to total dissolved material for different
sectors from which the air
masses were coming.

As can be seen the pH values as well as the acid fractions
do not vary very much. It is most astonishing that in cases
where the air had no land contact, i.e. transport direct
over the North Sea the acid fraction with 18% is only less
than half as when the air is transported over the Ruhr
area (24%). If the city or the industry of Hamburg would
have a marked influence this difference between these two
sectors should be even more pronounced. The small difference,
however, means that either acidification proceeds very
rapidly also at low SO_2 concentrations or that the 850 mbar
trajectory does not describe the transport of air pollutants
good enough i.e. that though no land contact is indicated
there is a marked cross-trajectory flow below 850 mbar
with continuous mixing upwards.

4. RESULTS OF CASE STUDIES

Case studies are suitable for proofing of concepts or a
hypothesis. Here we test with a case study the influence
of drops size spectra on the trace substance content. Fig 5
shows the weather pattern of the 20.7.79 where Hamburg
remained in a uniform airmass with several showers.

Two time periods with about the same rain intensities but
one with increasing and one with decreasing electrical
conductivity have been selected to study the influence of
rain drop size spectra. The rain drop size spectra were
measured with the disrometer after Joss and Waldvogel
(1967). As can be seen from fig 7 the presence of small
droplets increases the conductivity i.e. the trace

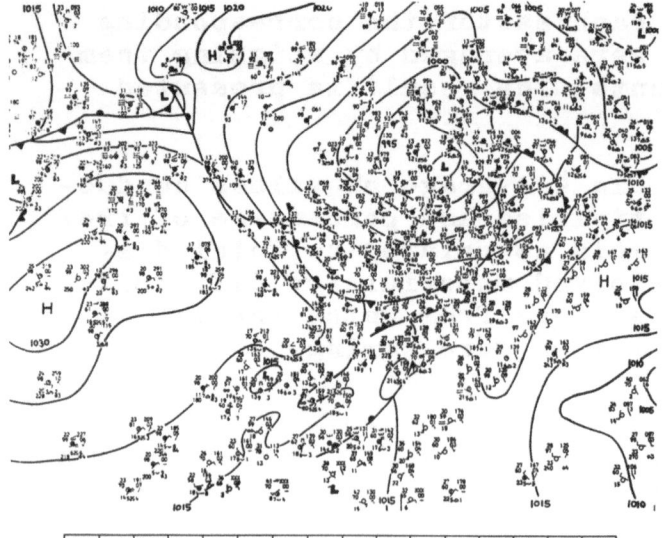

Fig 5 Weather situation
of the 20.7.1979

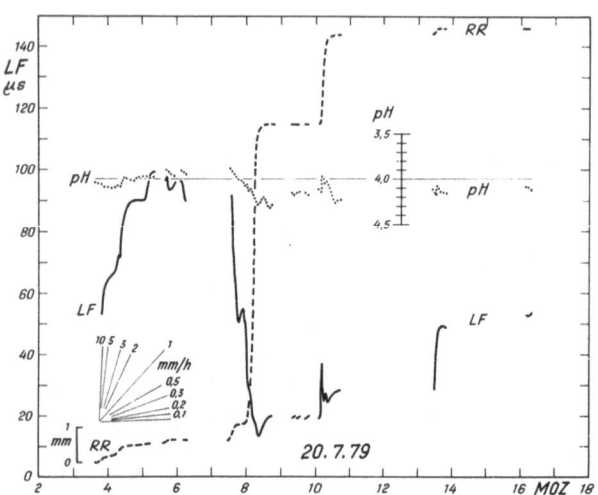

Fig 6 Record of pH
(dotted lines) electrical
eonductivities (full
lines) and precipitation
amount (RR, dashed lines)
of 20.7.79 during sereval
showers. Time periods
with increasing (around
5.00) and with decreasing
(around 8.30) electrical
conductivities have been
evaluated for the influ-
ence of drop size spec-
tra.

substance content of the precipitation whereas the absence
of small droplets decreases the conductivity. It should be
mentioned that the distrometer does not see the very small
drops below 0.3 mm in diameter so that the spectra are
incomplete. If we calculate the total surface of all drops
giving the same precipitation amount we find that in the
case with increasing conductivity the surface is nearly
twice of that in the case with decreasing conductivity.

Another case study of the 7. and 8. of July 1978 relates
to the development of the free acid fraction within 5
single showers. Fig 8 shows again the records of pH, conductivity and precipitation amount. The showers 1 to 4 are of
low conductivity i.e. low trace substance content; the 5th
shower is dirty, i.e. the conductivity is three to four
times higher as in the earlier showers. In fig 9 the traces
in the pH-conductivity diagram are depicted. These traces

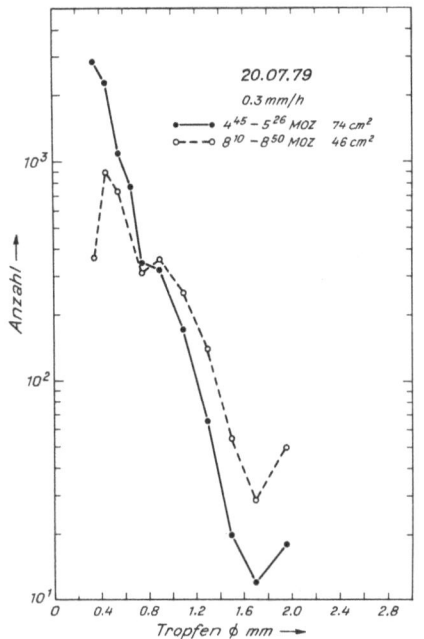

Fig 7 Rain drop size spectra for
two time periods of the 20.7.79.
The two figures 74 resp 46 cm^2
indicate the total surface as obtained by integration. The full
line belongs to a time period with
increasing conductivity, the dashed
line belongs to a time period with
decreasing conductivity.

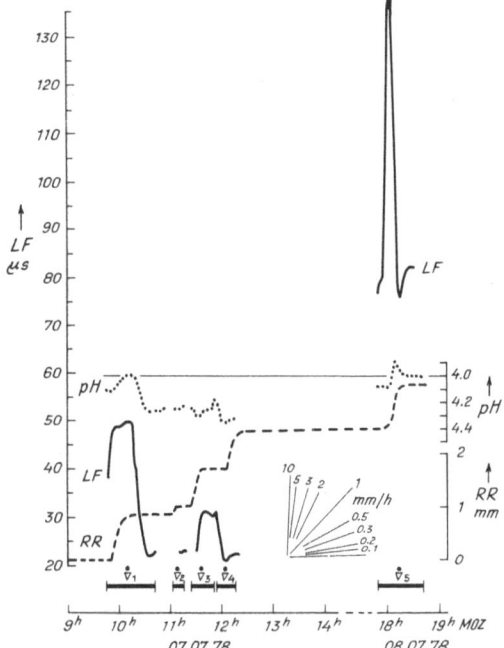

Fig 8 Record of pH (dotted lines)
electrical conductivities (full
lines) and precipitation amount
(RR, dashed lines) of 7. and 8.7.
1978

were constructed as follows. Every time when the tipping
bucket of the precipitation monitor gave a signal the pH
and conductivity were punched out on papertape. After
correction for the signal delay between the conductivity
cell and the pH probe these values were plotted into the
diagram.

As can be seen from fig 8 the acid fraction can increase
or decrease during a shower so that we cannot speak of a
principle behavior. Possibly the rain drops size distribution plays an important role.

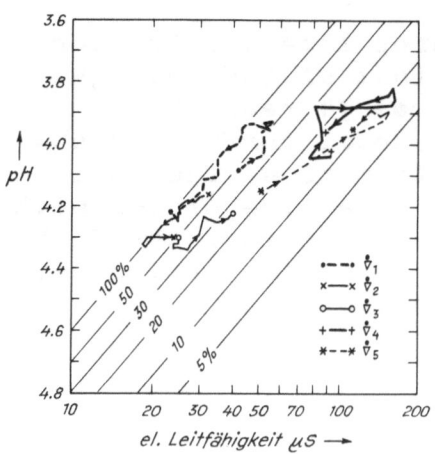

Fig 9 Traces of the acid fraction during 5 showers from the 7. and 8.7.1978

5. LONG TIME TREND

An important question is the development of the pH of precipitation in the last decades. From Scandinavia a decrease since the midfiftieth has been reported (Oden 1976) Although the collection was not free from influences due to dry deposition it seems to be doubtless that the decrease of rain pH has taken place (see fig 10).

A literature review shows that there exist measurements of rain pH from 1937 and 1938 of two stations (Reinerz and Oberschreiberau in Central Europe, ca 150 km northeast of Prague). As far as we can see these measurements have been performed carefully so that they are reliable. (Ernst (1938), Drischel (1940)). Because precipitation amount and pH values were published we can calculate a real weighed average value of pH 4.2 (see fig 10). Mrose (1966) reported of pH measurements from the Wahnsdorf Observatory near Dresden which were started in 1958. The average values are also near 4.2. Heigel (1960) described measurements at the Hohenpeissenberg observatory southwest of Munich. Unfortunately only frequency distributions were reported with highest frequency in the range 4.5-5.5 indicated as bar in fig 10.

Landsberg (1954) described pH measurements of single drops performed in Boston. An average value of 4 can be estimated. We have no reason that this value should not be transferable to european cities. For reasons of comparison we have plotted into fig 10 the pH development of two scandinavian cities and of Hamburg. We see that in Central Europe the pH of the precipitation cannot have changed very much. What we can conclude in connection with the rain characteristic in chapter 3 is (1) that in Central Europe the point where precipitation is dominated by acid was already

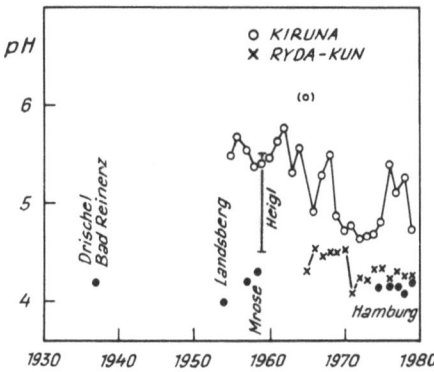

Fig 10 Long time trend of the
pH development in Central Europe
in the last five decades.

reached before 1937 (2) that further acidification proceeds
more slowly because due to the limited life time of precipitation elements the acid acumulation in a precipitation
is also limited (3) that as a consequence of (2) a precipitation event cannot remove all acidifying gases and
finally (4) that increasing emissions will not acidify the
precipitation near the source beyond a certain point but
precipitation is acidified over a larger area. This last
point is in agreement with the observations in Europe and
in North America.

6. CONCLUSIONS

Rain acidity is caused only to a small degree by acid
aerosol particles. Below pH 4.3 in polluted areas and below
pH 4.6 in less polluted areas the precipitation is dominated
by acid. At higher pH values the acid fraction lowers
rapidly.
Rain events do not remove acidifying gases completely.
Consequently an increase of acidifying emissions produces
not a rain whichis more acid in the neibourhood of the
emittors but when a certain pH (around 4) is reached the
area with acid precipitation is enlarged.

During the last five decades the rain pH in Central Europe
has not changed very much. This also demonstrates the
limited removal capability of the precipitation for
acidifying gases which have muliplied their concentration
several times since 1937. This limited removal capability
is balanced by long range transport.

LITERATURE

Drischel,H., "Chlorid- Sulfat- und Nitratgehalt der at-
mosphärischen Niederschläge in Bad Reinerz und Oberschei-
berau im Vergleich zu bisher bekannten Werten anderer Or-
te".
Der Balneologe 7 (1940) 321 - 334

Ernst,W., "Über pH-Wertmessungen von Niederschlägen"
Der Balneologe 5 (1938) 545 - 549

Granat,L., personal communication (1980)

Heigel,K., "Die pH-Werte von Niederschlägen, Kondensaten,
Nebelfrostablagerungen und der Schneedecke auf dem Hohenpeißenberg. Unterschiede der pH-Werte von Hohenpeißenberg
und Peißenberg".
PAGEOPH 47 (1960) 142 - 154

Joss,J. und A.Waldvogel, "Ein Spectrograph für Nieder-
schlagstropfen mit automatischer Auswertung"
PAGEOPH 68 (1967) 240 - 246

Landsberg,H., "Some observations of pH precipitation elements"
Arch.Met.Geoph.Bioclim. Ser. A 7 (1954) 219 - 226

Mrose,H., "Measurement of pH and chemical analysis of rain
snow and fog-water"
Tellus 18 (1966) 266 - 270

Oden,S., "The acid problem - an outline of concepts"
J.Air Water Soil Poll. 6 (1976) 137 - 166

Winkler,P., "Automatic analyser for pH and electrical conductivity of precipitation" in: Papers presented at the
WMO Technical Conference on Instruments and Methods of Observation (TECIMO) Hamburg 1977. WMO Nr. 480, Geneva 1977,
pp 191 - 196

COMPOSITION OF ACID RAIN IN THE FEDERAL REPUBLIC OF GERMANY
- SPATIAL AND TEMPORAL VARIATIONS DURING THE PERIOD
  1979-1981

Cornelia Perseke
Institute of Meteorology and Geophysics,
University of Frankfurt/M

ABSTRACT

An evaluation of the wet deposition measurements,
carried out on a regional scale in the FRG during the
period 1979-1981 is presented for pH, sulfate and nitrate.
pH values are observed in the range of pH 4.0-4.5.
The temporal variation of the sulfate and nitrate concentration in rain seems to be similar at all stations showing
maximum values during spring. The regional sulfate and
nitrate concentration patterns in rain differ between polluted and less polluted areas by a factor 2-3. The regional distribution of the wet sulfate and nitrate deposition
is mainly determined by the precipitation pattern.
The relative importance of sulfate, nitrate and chloride
for the acidity in rain is discussed.

INTRODUCTION

In the scope of a project, sponsored by the Umweltbundesamt a deposition network was in operation in the
Federal Republic of Germany during the period 1979-1981.
One main object of this undertaking were precipitation
chemistry measurements in order to assess wet deposition
on a regional scale. The study of the phenomenon "acid
rain" with pH values between pH 4-4.5, which had been observed in wide areas of the United States and Western Europe
was of special interest (Georgii 1981, Wilson et al. 1980).
Recently the problems connected with "acid rain" have gained growing attention because of some serious effects to
aquatic and terrestic ecosystems.

H.-W. Georgii and J. Pankrath (eds.), Deposition of Atmospheric Pollutants, 77–86.
Copyright © 1982 by D. Reidel Publishing Company.

MEASUREMENTS

The German deposition network, consisting of 13 stations, was established in cooperation with the German Weather Service (Fig. 1).

Fig. 1:   The German deposition network

At 10 stations the rain samples were collected over a full 2 year period from summer 1979 to summer 1981. The stations were located in polluted areas like the Ruhr area, the Rhein Main area and Hamburg. Less polluted areas were represented by the station Deuselbach/Hunsrück and by mountain sites (Kl. Feldberg/Taunus, Hoher Peißenberg/Upper

Bavaria, Schauinsland/Black Forest. At all stations rain
was sampled on a daily basis by means of a wet/dry collector to separate wet and dry deposition (Rohbock 1981). The
chemical analyses of the rain samples for the ions sulfate,
nitrate, chloride were carried out at the Nuclear Research
Establishment Jülich. For the sulfate analysis a comparison between ionchromatography and the isotopic dilution
method, according to Klockow et al. (1974) shows a good
agreement within 10%.

RESULTS AND DISCUSSION

In a recent paper by C. Perseke et al. (1981) first results of the deposition measurements were presented, which
are herewith updated.

ACIDITY

The temporal variation of the monthly averaged pH-values
during the period October 1979 to August 1981 is presented
for 4 stations (Fig. 2).

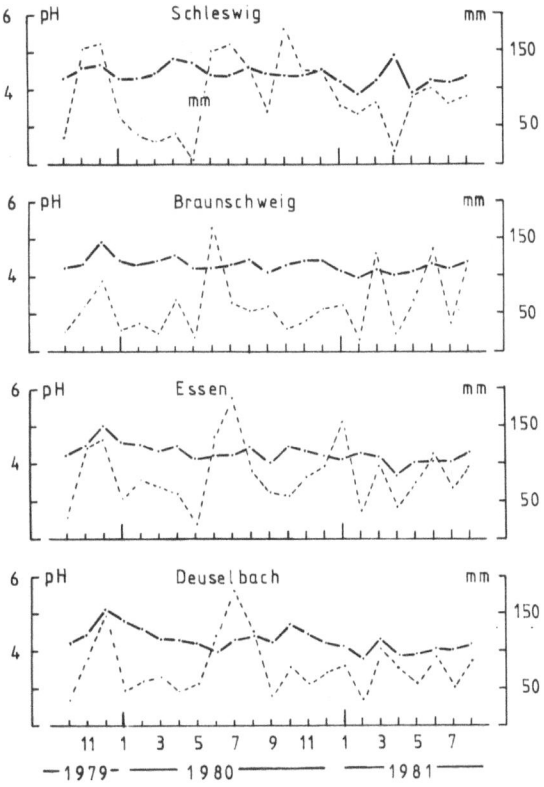

Fig. 2: Temporal variation of pH in rain during
         the period 1979-1981

For comparison the monthly precipitation rates are plotted, too. It can be seen that the variation seems to be
similar at all stations. Low pH-values, indicating high
acidity, are observed from January to March 1980, during
June and July and during September and October. At the
station Deuselbach the variation of pH is somewhat different. Whereas during the first year (1979) a decrease
of pH is observed from December 1979 until June 1980,
during the next year the decrease is found from October
1980 to February 1981. A comparison with the monthly precipitation rates shows no clear relationship. During autumn the increase of pH-values is found with an increase
of the rainfall amount. Whereas during summer 1980 low
pH-values are observed, although a long lasting precipitation period occurs.

The range of pH values and the difference between polluted stations (Frankfurt) and less polluted stations
(Deuselbach, Hof) is to be seen by a cumulative frequency
distribution (Fig. 3).

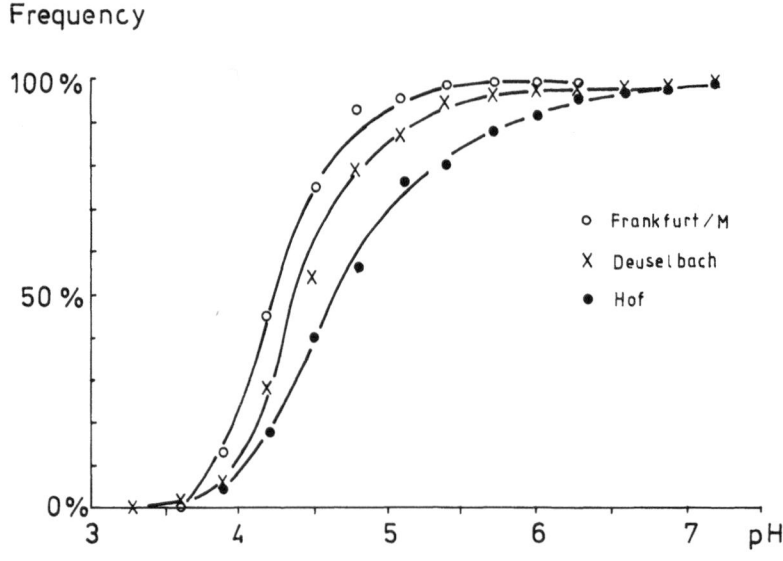

Fig. 3: Cumulative frequency distribution of
pH in rain

The frequency distribution shows the wide range of pH-values between 3.6 and 6.5-7.0, corresponding to a range in
H^+ by three orders of magnitude. Lowest pH-values occur
in Frankfurt, where 50% of the rain samples during the
period August 1979 to September 1980 show pH values lower
than 4.2. For the less polluted stations Deuselbach and

Hof the frequency distribution is shifted to higher pH values (50%-values $<$ pH 4.4 at Deuselbach and $<$ pH 4.6 at
Hof). The difference in pH between Frankfurt and Hof is
equivalent to a difference in H^+-ions by a factor more than
2.5. This is interesting because similar differences between polluted and less polluted areas are found for the
sulfate concentration, which gives further indication on
the relationship between pH and sulfate concentration in
rain. Klockow et al. (1978) found a good correlation between H^+ concentration versus the excess sulfate concentration ($SO_{\overline{4}}$ - Ca^{++}), thus showing that the sulfate plays an
important role for the acidity in rain. Investigations performed in the scope of the OECD project "long range transport of air pollutants" (Ottar 1978) and measurements in
the United States (Pack 1979) revealed that the main acid
producing components are sulfate, nitrate and chloride.
An evaluation of the relative contribution of sulfate, nitrate and chloride with respect to the acidity in rain was
carried out for yearly depositions, assuming for each mole
of sulfate an equivalent of two moles H^+ and for each mole
of nitrate, resp. chloride an equivalent of one mole H^+,
respectively (Fig. 4).

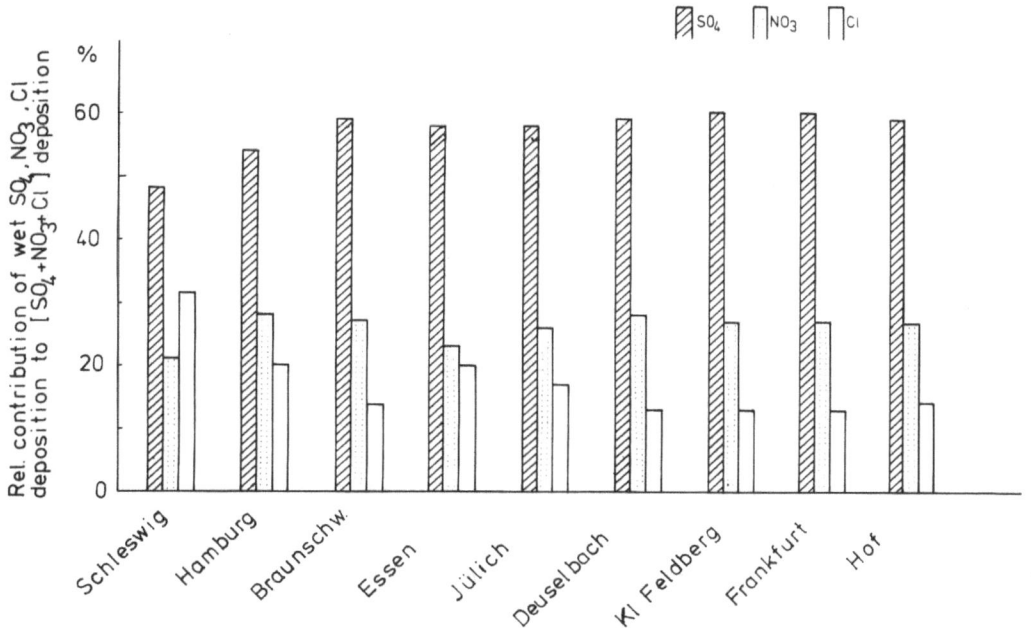

Fig. 4:   Relative contribution of the wet SO_4,
          NO_3, Cl deposition to total wet deposi-
          tion (units Val %)

With the exception of the stations nearby the coast, the
ionic composition of wet deposition is relative constant.
It can be seen that the wet SO_4-S deposition makes up
50-60% of the total wet deposition. The wet NO_3-N deposition contributes up to 25-30%. The chloride deposition
is increased nearby the coast and in the highly industrialized Ruhr area. Whereas at the other stations, the chloride deposition contributes less than 15% to the total wet
deposition. These values agree quite well with measurements in the United States (Wilson et al. 1980).
It was of interest to look for the temporal trends of the
acid producing components sulfate and nitrate.

TEMPORAL VARIATION OF THE SULFATE AND NITRATE CONCENTRATION
IN RAIN

The temporal variation of the sulfate concentration in
rain (weekly averaged values) is presented for the stations Essen and Deuselbach during the period 1979-1981
(Fig. 5).

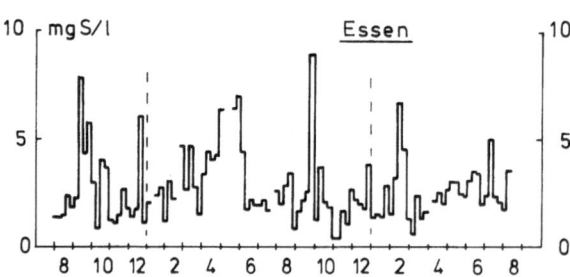

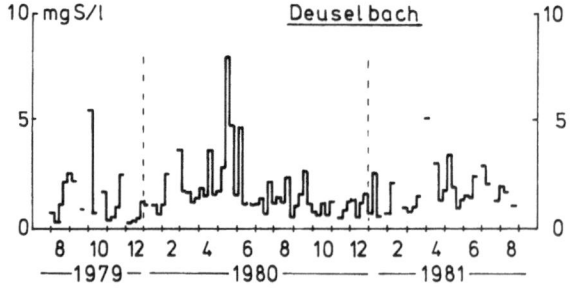

Fig. 5: Temporal variation of SO_4-S concentration
         in rain during the period 1979-1981

The station Essen is representative for a boundary region
of the heavy industrialized Ruhr area. Whereas Deuselbach
is situated in an unpolluted area in the Hunsrück-mountain.
It is to be seen that the sulfate concentration is quite
variable. The variability is the result of different influences, like precipitation rate and amount, winddirection
and general weather situations. Especially high concentration values are found with precipitation amount about or
less than 1-2 mm. The duration, resp. the total precipitation amount of a rain event is important for the final concentration of the rain samples (Kins 1981).
During autumn 1979 the sulfate concentration decreases to
values between 1 mg S/l in the polluted area Essen and O.2
mg S/l in less polluted areas (Deuselbach) in November and
December. High sulfate concentration values (8 mgS/l) are
observed during May 1980. Whereas at Essen further peak
values occur during September 1980 and February 1981 the
maximum during May is clearly pronounced in Deuselbach. In
1981 there is an indication of increasing sulfate concentration during spring. In order to explain the high concentration values in May 1980 the general weather situation
was taken into account. The weather situation in May 1980
was characterized by a weak gradient with an anticyclonic
flow pattern, resulting in a dry period of ten days. Precipitation with elevated concentration values occured before and after this dry period. The precipitation events,
mostly in the form of showers, were associated with post-
frontal advection of cold air masses. Before the dry period continental airmasses of polar origin were advected
from East Europe. Whereas the high concentration values
after the dry period are found with aged maritim polar airmasses, reaching Germany from southwest. Low concentration
values in June and July 1980 are the result of a long
lasting west weather situation with frontal precipitation
systems crossing Germany in short intervals.

The nitrate concentration in rain shows a similar temporal variation as the sulfate concentration. The nitrate
values vary between O.1 - 7 mg N/l at Essen and between
O.5 - 1 mg N/l at Deuselbach. Only from February to May
the nitrates are increased at Deuselbach.

REGIONAL DISTRIBUTION OF THE SULFATE AND NITRATE CONCENTRA-
TION IN RAIN

The regional distribution of the sulfate concentration
in rain shows a maximum in the highly industrialized Ruhr
area with values of 1.9 mg S/l during winter decreasing to
values of O.7 mg S/l in less polluted areas. During sum-

mer higher sulfate concentrations are observed at all stations. The gradient from polluted to less polluted areas
is not as pronounced as in winter. The rather uniform
pattern of the sulfate concentration in rain indicates
that wet deposition of acid substances is an interregional
problem. By transport of polluted airmasses and incorporation into frontal precipitation systems "less polluted"
areas receive precipitation with a high content of acid
substances.

    The regional distribution of the nitrate concentration
in rain shows differences with regard to the maximum area.
During winter highest nitrate concentrations of 0.7-0.8 mg
N/l occur further northeast in the region of Hamburg and
Braunschweig, decreasing to values of 0.2 mg N/l in Deuselbach. During summer the pattern changed. The maximum area
with values of 0.85 mg N/l is located southwest of the
Ruhr area. At the northern stations lower nitrate concentrations are observed in summer compared to winter. Whereas at the southern stations, esp. Deuselbach higher concentrations are found in summer.

WET DEPOSITION

    On the basis of the concentration values and precipi-
tation amounts, the wet deposition was assessed. The knowledge of the wet deposition is necessary for the valuation
of effects. The wet SO_4-S deposition is increased during
the summer season due to a long lasting precipitation
period in June and July 1980 (Fig. 6).

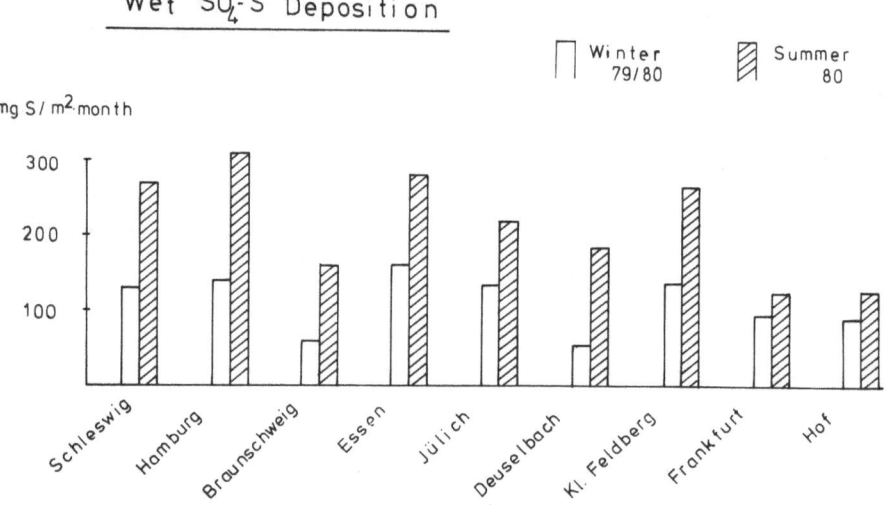

Fig. 6: Wet SO4-S deposition during winter and
         summer season

Maximum wet sulfur deposition values are found in the highly industrialized Ruhr area and in areas with high precipitation amounts like mountain stations. A comparison with the pattern of total rainfall amount makes obvious that wet deposition is increased in areas with enhanced rainfall amounts. The highest percentage of wet deposition is found with advection from southwest to west (at surface layer) as a result of the prevailing southwest-westwinds during frontal precipitation.

CONCLUSIONS

The measurements of wet deposition emphasize, that acid precipitation with increased sulfate and nitrate concentrations occurs in heavy polluted areas as well as in less polluted areas. In less polluted areas outside the emission areas the concentrations of the acid substances in rain is reduced by a factor 2-3, only. Furthermore, the wet deposition pattern is mainly determined by the precipitation pattern. This is of main importance for less polluted mountain regions which receive high wet depositions of acid substances.

ACKNOWLEDGEMENT

This investigation was sponsored by the Umweltbundesamt under contract No. 104 02 600. We want to thank the staff of the Deutscher Wetterdienst, the Geophysikalischer Beratungsdienst, and the Umweltbundesamt for their technical assistance by collecting deposition samples.

Special thanks go to Mrs. G. Aheimer and Mr.K.P. Müller, Institut für Atmosphärische Chemie, Kernforschungsanlage Jülich GmbH, who performed the analyses by ionchromatography. The staff of the laboratory of the Institut für Meteorologie und Geophysik, Frankfurt/Main, especially Mrs. M. Obeth is thanked for their accurate rainwater analyses.

REFERENCES

Georgii, H.-W. (1981) Review of the acidity of preci-
    pitation according to the WMO-network.
    IDŐJÁRÁS Vol 85, 1-9

Kins, L. (1981) Temporal variations of the chemical
    composition of rainwater during single precipita-
    tion events - measurements and interpretations, in:
    Deposition of Atmospheric Pollutants, Ed. H.-W. Georgii,
    J. Pankrath D. Reidel Publishing Company, p. 87.

Klockow, D., Denzinger, H., Rönicke, G. (1974) Anwendung
    der substöchiometrischen Isotopenverdünnungsanalyse
    auf die Bestimmung von atmosphärischem Sulfat und
    Chlorid "Background"-Luft.
    Chemie Ing. Techn. 46, Heft 19

Klockow, D., Denzinger, H., Rönicke, G. (1978) Zum Zu-
    sammenhang zwischen pH-Wert und Elektrolyt-Zusammen-
    setzung von Niederschlag.
    VDI Berichte Nr. 314, pp 21-26

Ottar, B. (1978) An assessment of the OECD-study of long
    range transport of air pollutants.
    Atm. Env. 12, 445-454

Pack, D. H. (1979) Acid precipitation - the physical
    system.
    Proc. Advisory Workshop on the Formation of Acid Pre-
    cipitation, EPRI Rep. EA 1074, 3-137

Perseke, C., Georgii, H.-W., Rohbock, E. (1981) Investiga-
    tion of the regional distribution of wet deposition of
    pollutants.
    Proceedings of Second European Symposium, 29 Sept -
    1 Oct 1981, Varese

Rohbock, E. (1981) Removal of airborne heavy metals by
    dry and wet deposition, in: Deposition of Atmospheric
    Pollutants, Editors, H.-W. Georgii, J. Pankrath,
    D. Reidel Publishing Company, p. 159.

Wilson, J., Mohnen, V., Kadlecek, J. (1980) Wet deposition
    in the Northeastern United States.
    ASRC Publication 796, State University of NY, Albany

TEMPORAL VARIATION OF CHEMICAL COMPOSITION OF RAINWATER DURING INDIVIDUAL PRECIPITATION EVENTS

Lucia Kins
Institute for Meteorology and Geophysics
Frankfurt/Main

ABSTRACT

An automatic sequential precipitation sampler was developed and constructed. It collects rainwater on a volume basis. The sample volume can be selected for each precipitation event due to the minimum for the analysis. The sampler is working since July 80. Analysis includes pH, conductivity, lead, manganese and sulphate. In general the concentrations of trace elements are high at the beginning of a rainfall, they decrease rapidly and remain at a constant value. For convective showers rainfall intensity and concentrations show strong inverse relationship. During rain events of longer duration, such as warm front precipitations, the trends of concentrations depend on the synoptic conditions.

INTRODUCTION

Sequential sampling and chemical analysis of rainwater provides more information about the scavenging of trace substances than bulk sampling. In the past investigations on the temporal variations of the concentration of trace substances during individual rainevents were mainly restricted to radioactive bomb debris (Kruger and Hosler 1963, Huff and Stout 1968, Noyce et al. 1971). They all found large concentration changes, especially during convective showers. The present investigation was performed to study characteristic concentration patterns of different precipitation types. For this purpose a sequential sampler was constructed.

H.-W. Georgii and J. Pankrath (eds.), Deposition of Atmospheric Pollutants, 87–96.

SEQUENTIAL SAMPLER

The collector operates on the basis of equal samples
with a minimum volume necessary for chemical analysis.

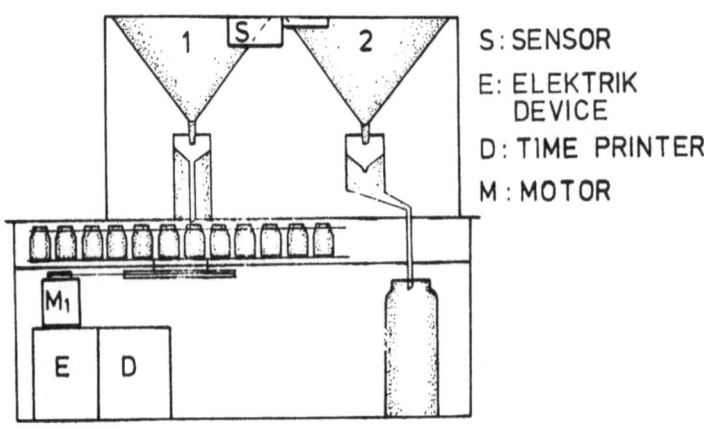

S: SENSOR
E: ELEKTRIK
   DEVICE
D: TIME PRINTER
M: MOTOR

Fig. 1: Sequential sampler

The collector consists of two equal funnels made of polyethylene (Fig. 1). Both are covered during dry periods.
With the beginning of a rainfall a sensor initiates the
opening of the lid. The rain samples are divided into
droplets of known and equal size when leaving the funnel.
They are counted by means of a photocell during their fall.
When the desired volume of the rain fraction is collected
the sampling dish moves to the next bottle. The required
number of droplets is selected at the beginning of each
precipitation event according to the volume required for
analysis.

A time-printer records the change of the sampling bottles,
and the opening and closing of the lid. Thus, the rainintensity during the precipitation event can be calculated.
The sampling dish is separated from the electrical and
mechanical device to avoid contamination of the samples.
For our purpose a sample volume of 15 ml is required, corresponding to a precipitation-rate of 0.36 mm. 48 bottles
are placed on the sampling dish, thus, a total amount of
rain of 17,4 mm can be collected.

ANALYSIS

The analysis of the samples is performed immediately
after collection.   Besides the measurements of conductivity and pH, sulfate is analysed by isotopic dilution
analysis (Klockow et al. 1974), lead and manganese by atomic absorption spectrometry.

MEASUREMENTS AND RESULTS

The individual rainevents are classified in three
types:
a)   convective showers
b)   convective precipitation events
c)   precipitation events connected with upslide motions.

CONVECTIVE SHOWERS

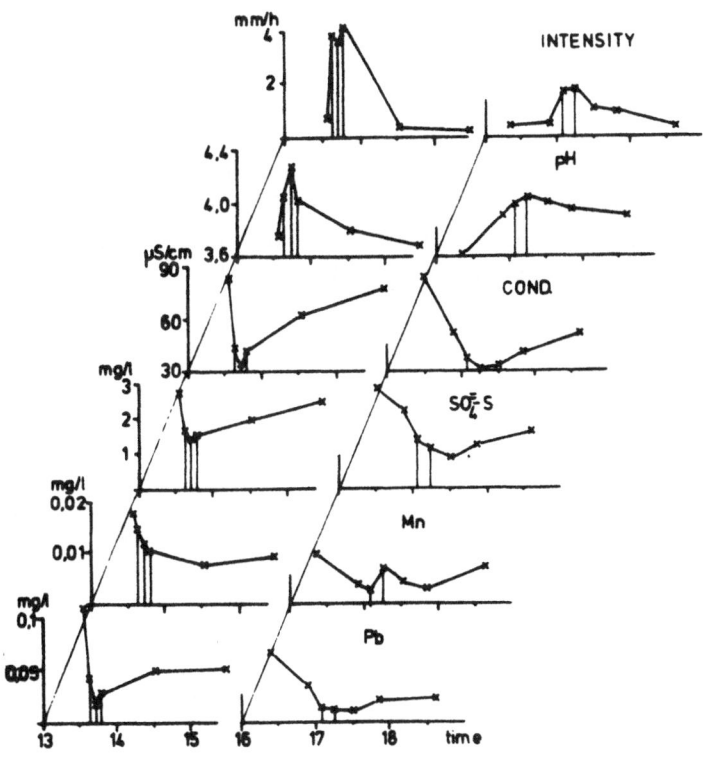

Fig. 2:   Rainintensity and concentration varia-
          tions during two convective showers

Fig. 2 gives an example of two convective showers. A
strong inverse relationship between the concentrations
and the rainfall rate is to be seen. This is in accordance
with previous investigations by Gatz and Dingle (1971).
The first shower is characterized by a sudden maximum of
intensity followed by light rain. The second shower had
only a less pronounced intensity-maximum. During the
first shower the value of conductivity and the concentrations of sulfate and lead show a clear minimum at the time
of the intensity maximum. The pH-value being inverse to
the conductivity, has a maximum at the same time. During
the light rain the sulfate concentrations and the conductivity increase up to the initial values.

During the second shower, only a slight concentration minimum of sulfate and lead and of the conductivity is found
just after the intensity maximum, pH-values have a minimum again. The decrease of these concentrations is slow
according to the slow increase of rainfall rate. Also
the increase of these concentrations after the most intensive rain is slower than during the first shower.
Manganese again shows no relationship between the variation of the concentration and the rainfall rate. It was
found that the concentration minimum is more significant
the greater the decrease of the rainfall rate and the higher
the intensity maximum.

Convective clouds with their large vertical extension incorporate trace elements by lateral entrainement. This
leads to a concentration gradient within the clouds. During
the time of the intensity maximum evaporation and washout
can be excluded because of high fall velocity of the drops
below the cloud-base. This explains the lower concentrations during the heaviest rain.

In contrast to the concentration-fluctuations of convective showers described above and their relation to the
rainfall rate, the concentration variations in cases of
precipitations with longer duration are less well documented. Georgii (1965) studied the variations of inorganic ions of some individual rainfalls as a function of
rain amount. Anderson and Landsberg (1979) showed variations of pH-values for different raintypes.

CONVECTIVE PRECIPITATION EVENTS

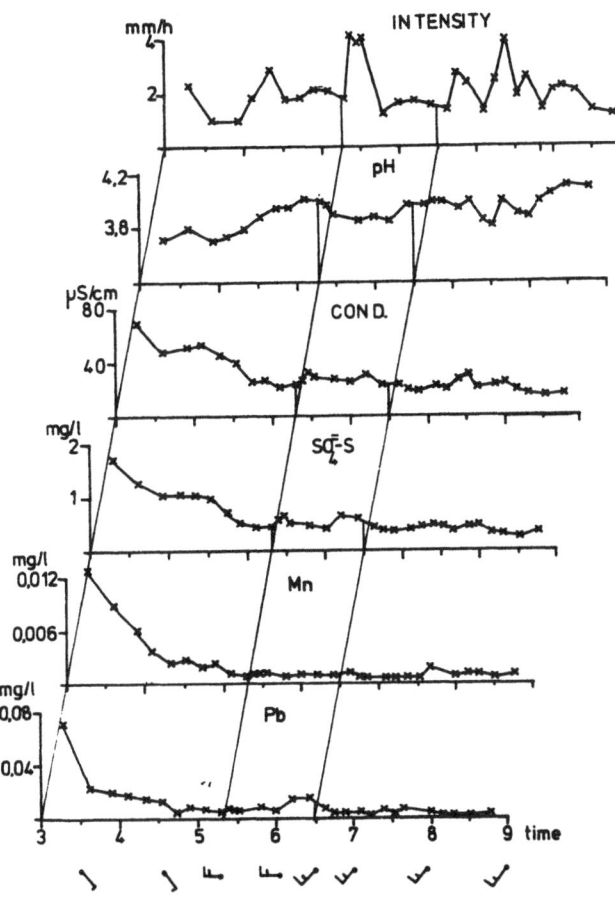

Fig. 3:   Concentration variations during a continous rain from a convective cloud system

Fig. 3 gives an example for a precipitating convective
cloud system.   During the rainfall the passage of a cold-
front was observed, it is characterized by the change in
winddirection and windspeed.   The sulfate concentration and
the conductivity show a sudden increase when the winddirection turned, while the pH decreased.   The concentrations of
lead and manganese do not show any remarkable change during
the frontpassage.   The decrease of sulfate concentrations
and the conductivity is clearly slower than for lead and
manganese.

Sulfate concentrations, conductivity and pH-values show
several small concentration maxima throughout the rainevent,
whereas the lead and manganese concentrations are more uniform.   The great variability of sulfate of conductivity
and pH-values is typical for precipitation from convective
clouds.   This can be explained by the dynamics of convective cloud systems consisting of many single cells and by
lateral entrainement.

PRECIPITATION CONNECTED WITH UPSLIDE-MOTION

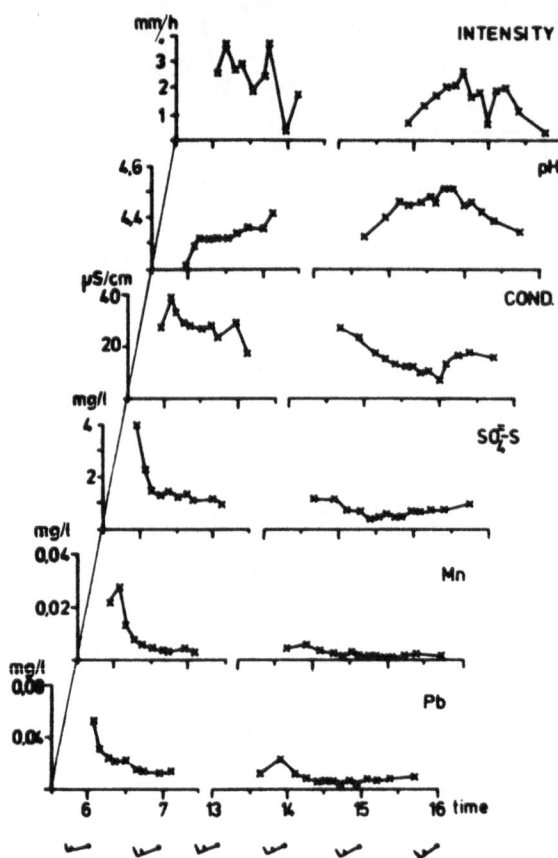

Fig. 4:   Concentration variations during a precipi-
          tation connected with upslide motion

Fig. 4 gives an example for continous rain. The rainfall
was interrupted by a dry spell of about six hours. During
the dry period no change of winddirection and windspeed
was observed. The trace-substance-concentrations in the
rainwater did not change during the dry period. This indicates that the rainfall originated entirely from the same
cloudsystem. At the beginning of the event, high concentrations of all elements are found, they are decreasing rapidly and remain quasi constant. The pH-value is inverse
to the conductivity.

The different behaviour of the concentrations of trace-sub-
stances in rain during convective and upslide precipitation
can be explained by the different cloud types. Upslide
movement prior to a warmfront leads to the formation of
stratiform clouds of great spatial extension. In contrast
to convective clouds, stratiform clouds are characterized
by a homogeneous droplet distribution, low vertical velocity, no lateral entrainement.

The examples presented in this paper have shown that every frontal precipitation event is connected with individual variations of the concentrations. However, the general trend (high initial concentrations, followed by a decrease and more or less constant values towards the end of the rain event) was found in every case.

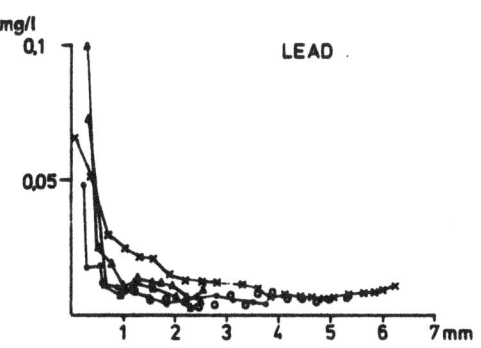

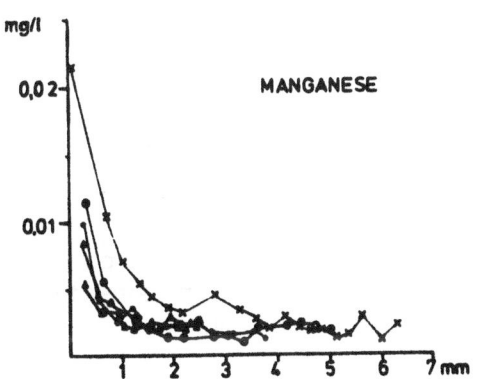

Fig. 5:   Lead and manganese concentration varia-
          tions for several rain events connected
          with upslide motion

In Fig. 5 the lead and manganese concentrations are plotted as a function of the amount of rainfall for cases connected with upslide-motion. It can be seen that rainfalls with more than 2 mm show no difference in the concentrations of trace substances. Only the initial concentrations vary.

In Fig. 6 the lead and manganese concentrations of several rainevents from convective clouds are presented. No signi-ficant differences to the concentration variations of up-slide precipitation are found. The initial concentrations again show the largest differences. Two main effects cause the values of the initial concentrations.

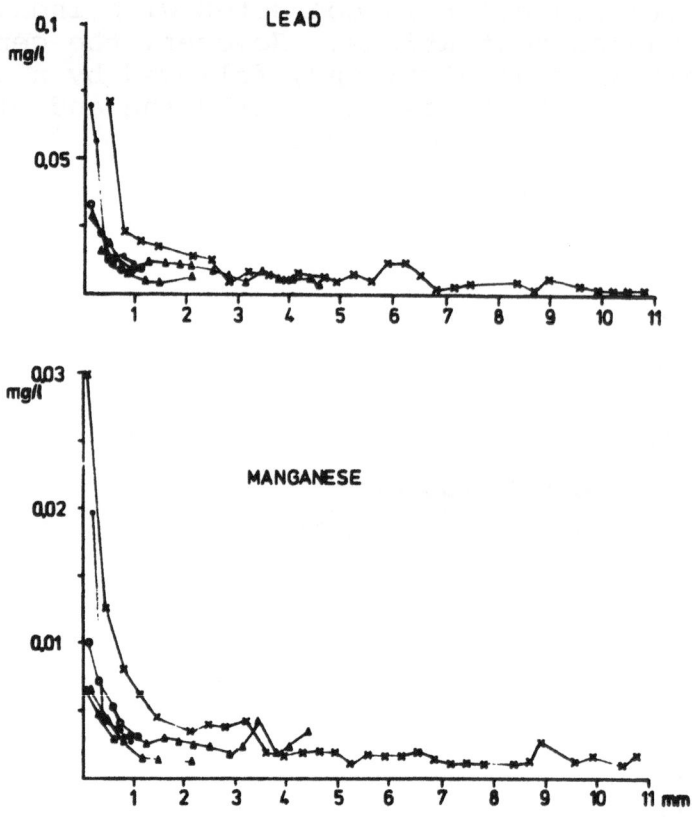

Fig. 6:   Lead and manganese concentration varia-
          tions for several rain events from con-
          vective cloud systems

a)   wash-out effect
b)   evaporation effect

Both figures show also that the decrease of the concentrations is a function of the rain amount and not of the
duration of rainfall. The sulfate concentrations and the
conductivity of the same precipitation events are presented
in Fig. 7 for continuous rain and in Fig. 8 for convective
rain. In contrast to lead and manganese the concentrations
of sulfate and the conductivity differ for each individual
rainfall. These large differences in the sulfate and the
conductivity of the individual rain events are mainly resulting from advective effects. Sulfate and other inorganic ions are advected to a larger extend with the air
masses while lead and manganese are produced predominantly
from local sources.

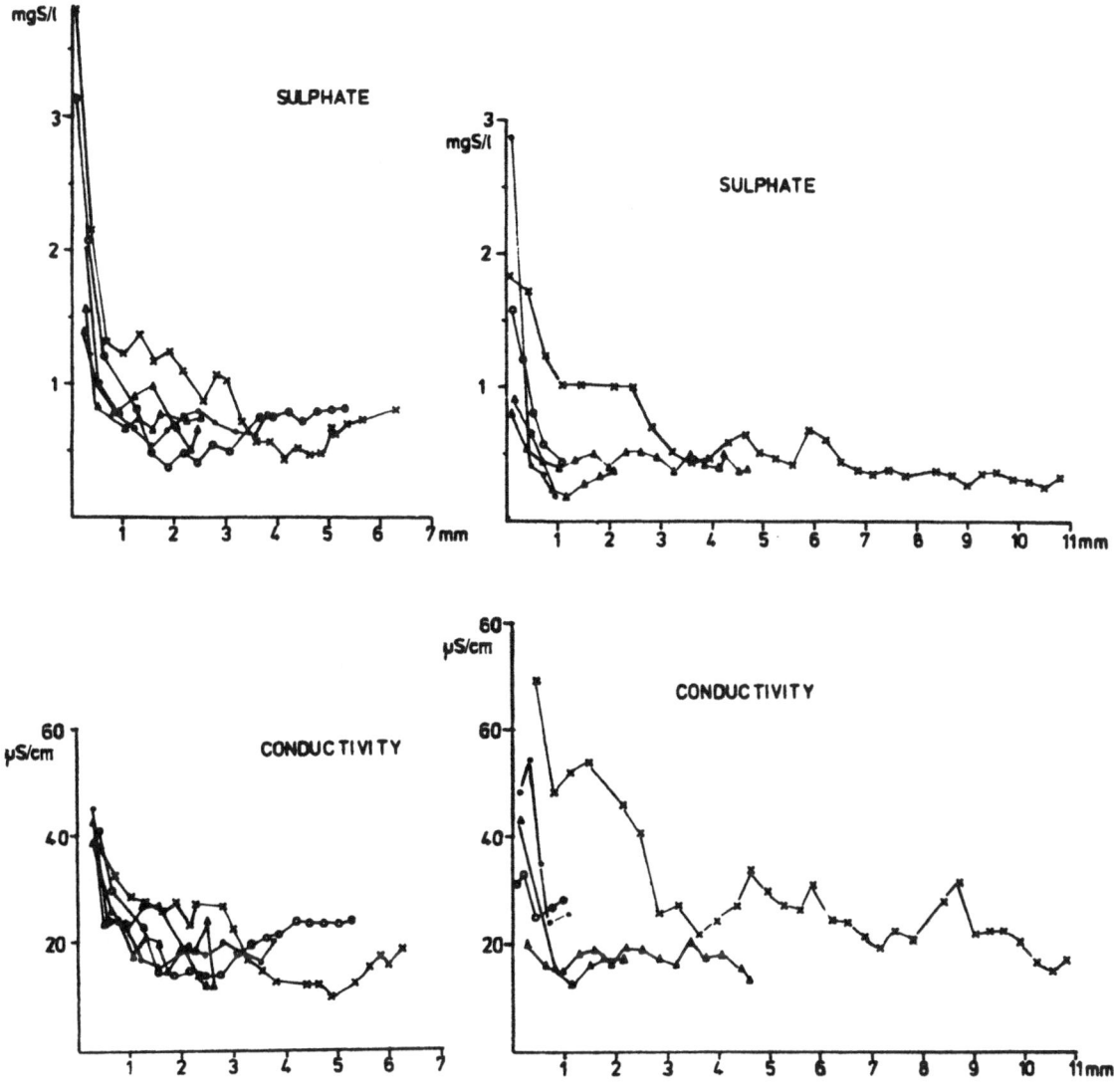

Fig.7: Sulfate concentration and conductivity variations for rainevents connected with upslide motion.

Fig. 8: Sulfate concentration and conductivity variations for rainevents from convective cloud systems.

CONCLUSIONS

The paper presents an automatic sequential precipitation sampler and first results of the fractional sampling of rain during individual showers and continuous rainfalls. The result can be summarized as follows.

1) Concentrations of convective showers show an inverse trend to variations.
2) pH and conductivity-values show contrary variations.
3) The concentrations of trace substances in convective precipitation shows larger variations during the rainfall compared to rain from stratiform clouds.
4) All types of precipitation show similar concentration-variations for lead and manganese. Constant values are reached after a rainfall-rate of two mm.
5) Variations of the sulfate concentration and the conductivity are different for every rainevent.

REFERENCES

Anderson, D.E. and Landsberg, H.E. (1979) Detailed structure of pH in hydrometeors.
Env. Science and Technology 13, 992-994

Gatz, D.F. and Dingle, A.N. (1971) Trace substances in rainwater: concentration variations during convective rains and their interpretation.
Tellus 23, 14-27

Georgii, H.-W. (1965) Untersuchungen über Ausregnen und Auswaschen atmosphärischer Spurenstoffe durch Wolken und Niederschlag.
Ber. des DWD 14

Huff, F.A. and Stout, G.E. (1968) Relation between ^{144}Ce and ^{90}Sr rainout in convective rainstorms.
Tellus 20, 82-87

Klockow, D., Denzinger, H., Rönicke, G. (1974) Anwendung der substöchiometrischen Isotopenverdünnungsanalyse auf die Bestimmung von atmosphärischem Sulfat und Chlorid in: Background-Luft.
Chemie Ing. Techn. 46, Heft 19

Kruger, P., Hosler, C.L. (1963) ^{90}Sr concentration in precipitation from convective showers.
J. Appl. Met. 2, 379-389

Noyce, J.R., Chen, T.S., Moore, D.T., Beck, J.N., Kuroda, P.K. (1971) Temporal distributions of radioactivity and ^{89}Sr/^{90}Sr ratios during rainstorms.
J. Geophys. Res. 76, 646-656

INVESTIGATIONS ABOUT WET DEPOSITION OF POLLUTANTS IN AN URBAN ECOSYSTEM

Wilhelm Kuttler
Geographisches Institut, Ruhr-Universität Bochum

1. Problems

In view of an extensive investigation programm on geo-ecological problems studies of acid precipitation were begun in the Ruhr District and its periphery at the following collection stations: Geldern (Lower Rhine, in the periphery of the Ruhr District (RD), Lünen (in the eastern part of the RD, Bochum (in the central area of the RD, Wuppertal (in the Bergisches Land), Lüdenscheid (in the Bergisches Land), Obergurgl (Tirol, Austria)
While the measurements at the stations of Geldern, Lünen, Wuppertal, Lüdenscheid and Obergurgl have been in process since early or mid 1980, and are intended to end in 1982, the measurements at Bochum station (Botanical Garden of Ruhr University) have still in process since May 1978. At this station the investigations are intended to run after 1982.
Precipitation is collected 1.6m above ground in bulk samplers, made of polyethylene, exposed for a week.Every Monday the samples are analysed among others on the pollutant compounds of sulphate, calcium, chloride as well as nitrate and lead. Additionally the pH-value is determined and since April, 1981, even electric conductivity.
Since October, 1981, we analyse parallel to these investigations event-only rain samples in order to find out the enrichment factor of concentrations in bulk samples to concentrations in pure samples (see MOLDAN, 1980, as well as GEORGII and others, 1980). Samples which cannot be analysed on the same day are kept in a refrigerator at + 4°C.
The methods of analysis are conventional and are carried out with the help of photometry and titration.

On the one hand it is the aim of the research programm to determine the amount of different pollutants, and on the other hand to examine the effects of acidity on soil (buffering capacity).

H.-W. Georgii and J. Pankrath (eds.), Deposition of Atmospheric Pollutants, 97–113.
Copyright © 1982 by D. Reidel Publishing Company.

Concerning this , extensive soil-dependent investigations
will be carried out in early summer 1982 in the Ruhr District as well as in the remaining NRW in order to map soil
sensitivity,similar to the investigations by ROOT and other
(1930) in the frame of a survey for North America and by
GLASS and others (1980) for the eastern coast of the United
States as well as HUTTUNEN (1979) for Sweden. Extensive
measurements concerning estimations of forest-ecosystems
have been carried out in the Federal Republic of Germany,
i.e. by ULRICH and others (1976) or GÜNTHER and others
(1976).
In this report I'll present the first results from Bochum
station for some selected pollutants, and that for calcium,
chloride and sulphate as well as the pH-value during the
period from May, 1978 to April, 1980.

2. Results
The frequency distribution of calcium concentrations at
Bochum station shows(Fig.1) values ranging between 0.5
mg Ca/l and 6.5 mg Ca/l. The mean value is 3.2 mg Ca/l.
Referring to this the average of winter months is lower
by 18 p.c. while the summer values are higher by 11 p.c. .
For Calcium is a significant component of dust and is used
especially in the building and construction trade it becomes
that the higher summer values can be interpreted as a result from the intense building activities during the summer months.
Turning to the chlorid concentrations (Fig.2) we could not
find a visible seasonal difference during the period of our
investigations. We found an average value of x = 4.4 mgCl/l.
The frequency distribution of the measured values ranged
between 1 and 15 mg Cl/l with a median value of 4.0 mg Cl/l.

The frequency distribution of the pH-values for the annual
mean shows (Fig.3) values between 3.2 and 7.0 at Bochum
station resulting in an average mean of 4.3. The seasonal
variations which show a relatively low value in the winter
period of pH = 4.0 with a summer value of pH = 4.5 can be
compared to the investigations of KAYSER and others (1974)
at "Schauinsland Station" in the Black Forest. This can be
explained together with KAYSER (1974) and KLOKOW (1978) in
the influence of the heating period.
Adequate to the importance of sulphur in our atmosphere I'll
deal with this pollutant more in detail.
The sulphate concentrations (Fig.4) reached an average of
x = 18.6 mg SO_4/l with a standard deviation of s = 6.3.
The lowest concentrations in rainwater had been measured
in July, 1979, with values between o.9 and 1.1 mg SO_4/l
when the situation of a cyclonal western weather condition
with maritime tropic air masses caused heavy rains and re-
sulted in a low absorption of pollutants in the raindrops.

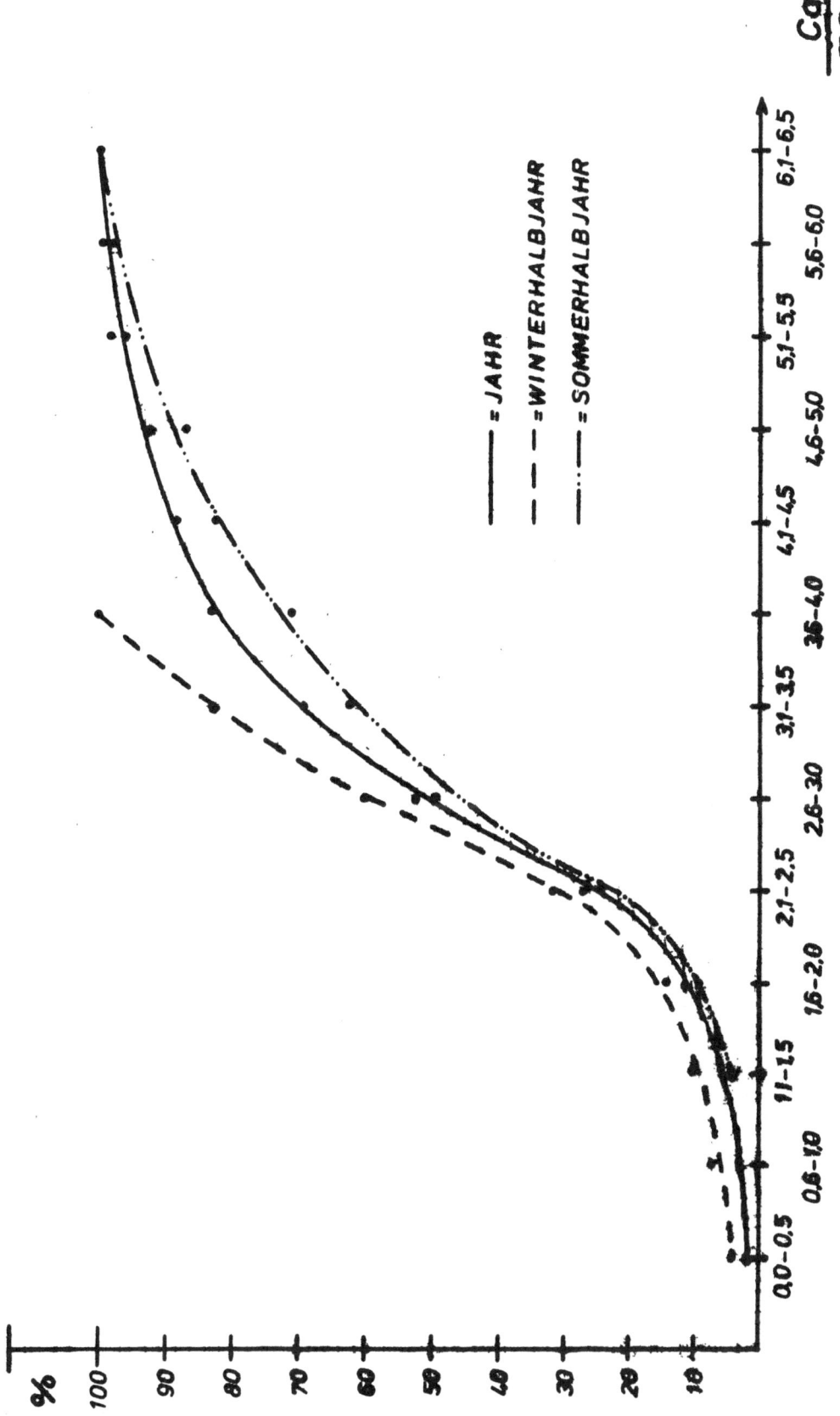

Fig. 1: Frequency distribution of calcium concentrations in precipitation

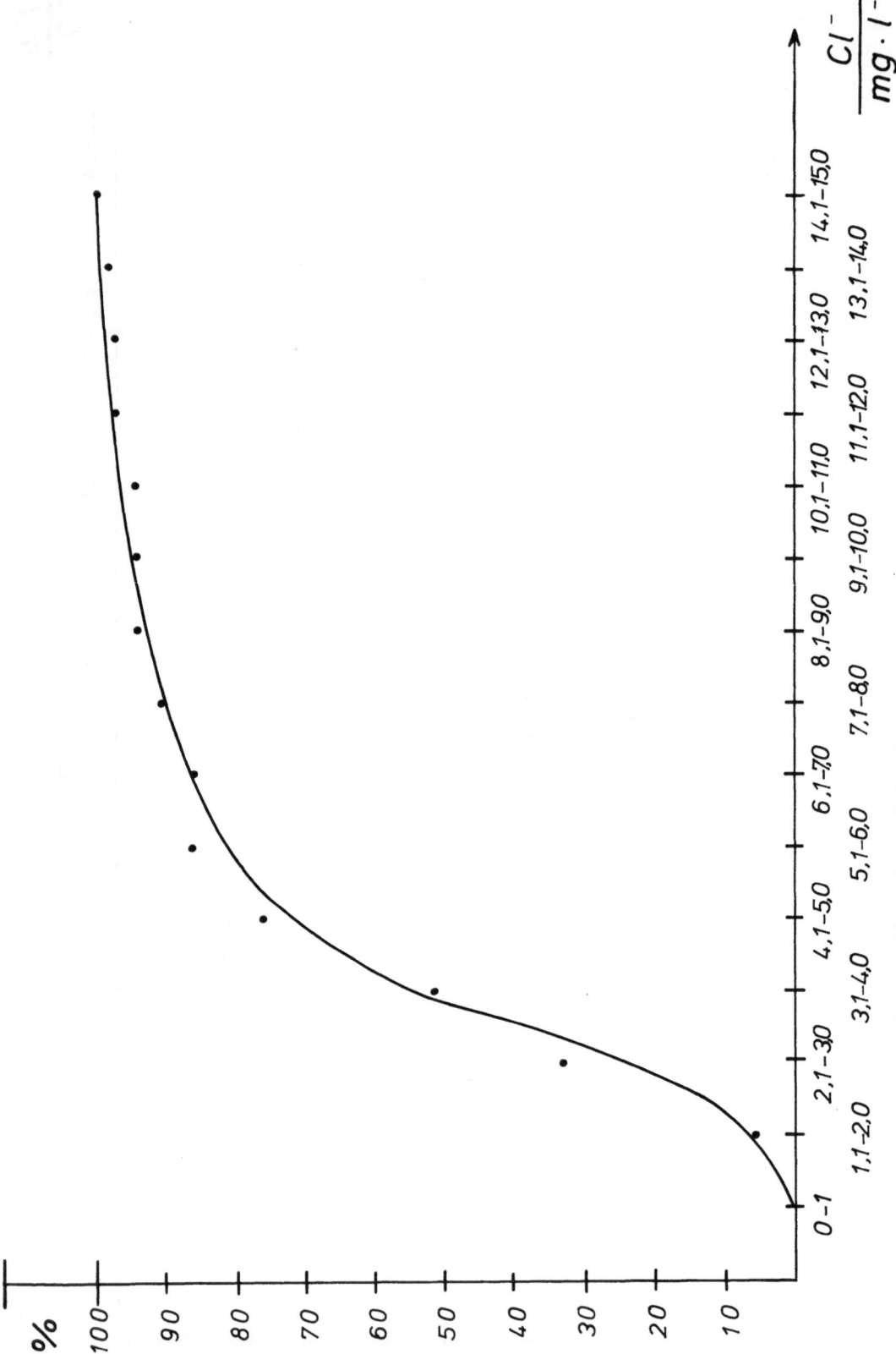

Fig. 2. Frequency distribution of chloride concentrations in precipitation.

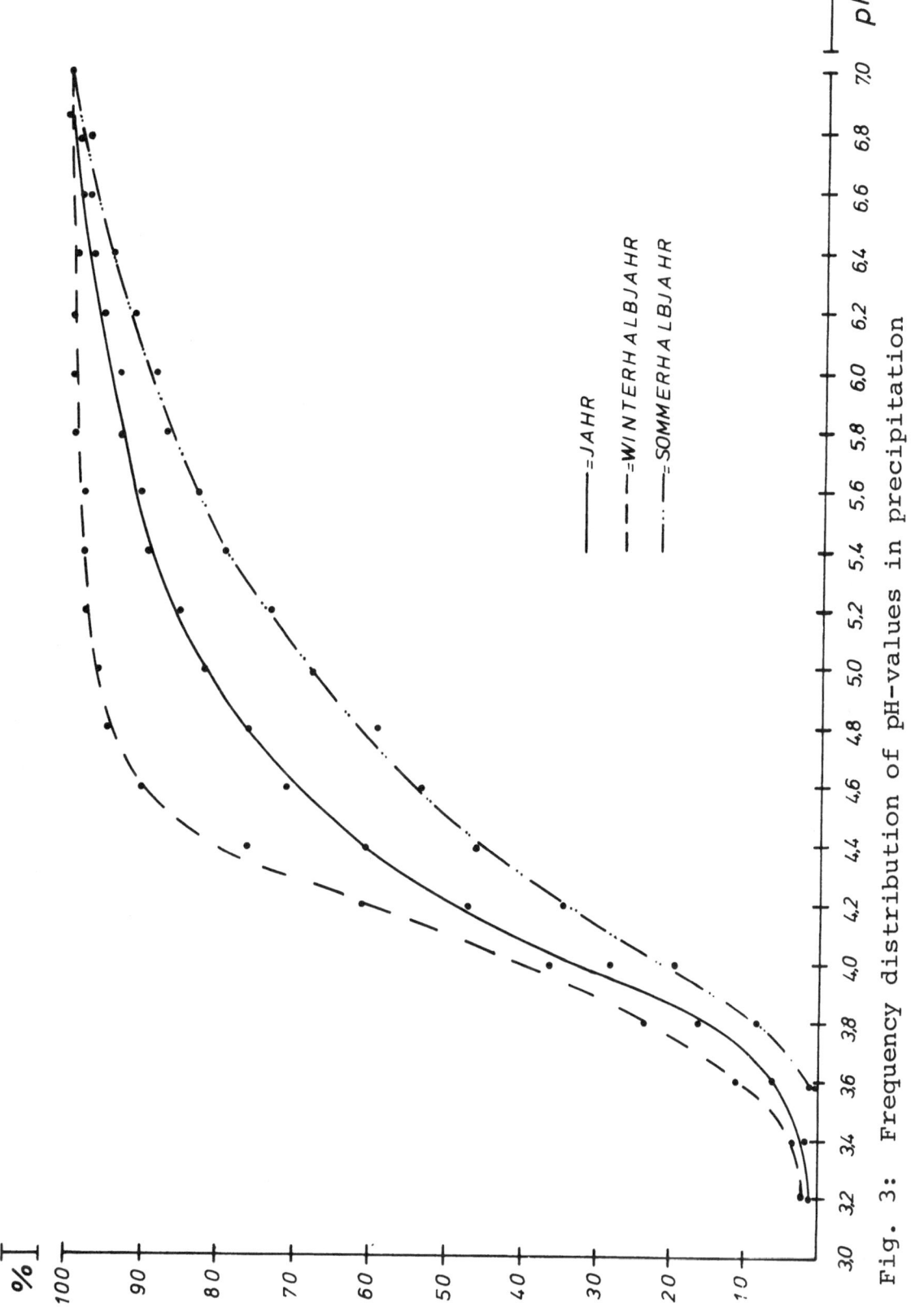

Fig. 3: Frequency distribution of pH-values in precipitation

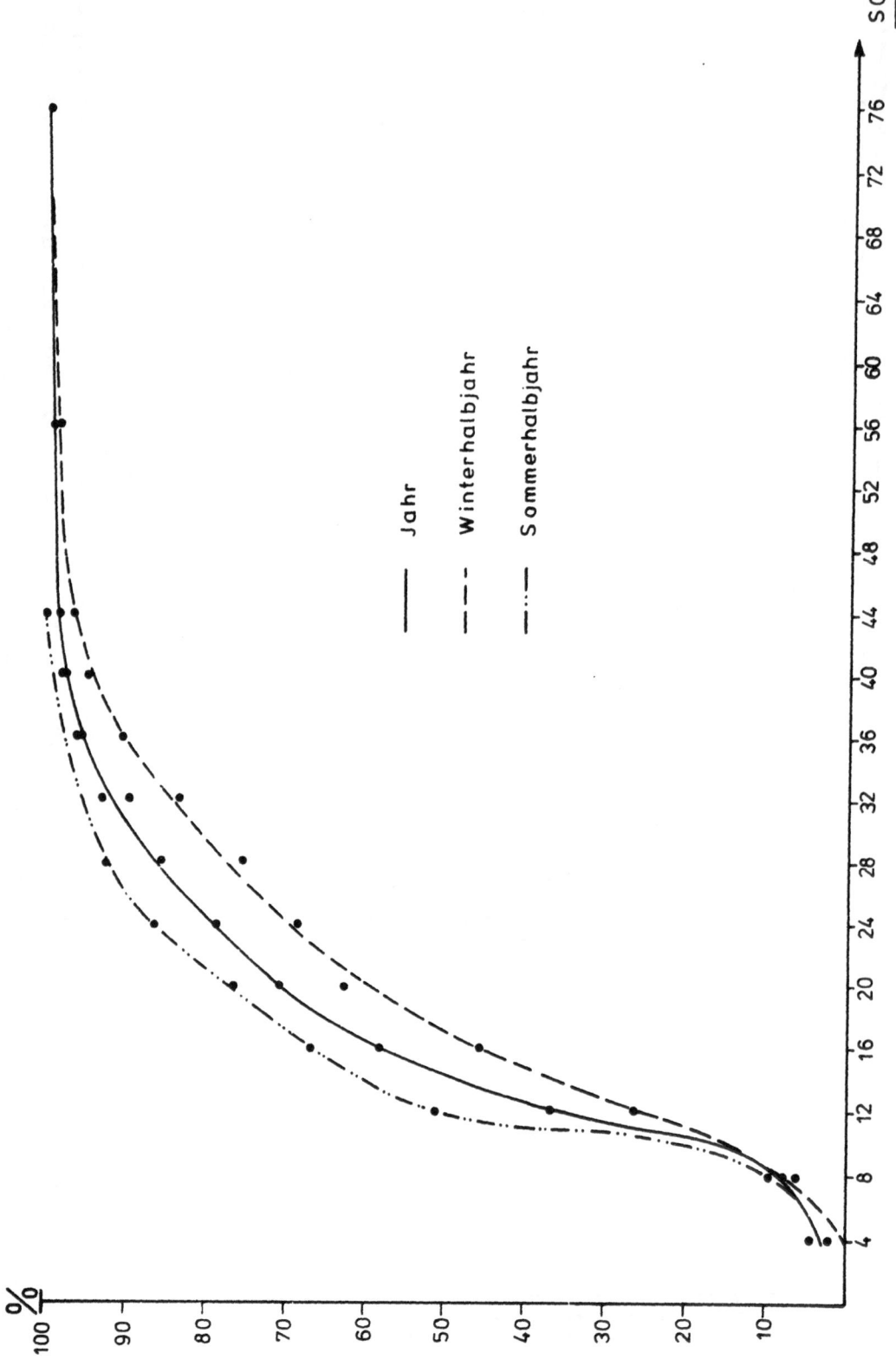

Fig. 4. Frequency distribution of sulphate concentrations in precipitation.

On the other hand the highest sulphate concentrations had
been registered in January, 1979, with 75.1 mg SO_4/l in the
first rain after anticyclonal weather conditions. During
this stable weather condition the concentrations of the
pollutant SO_2 had been high enough to develop a SO_2-Smog
(KUTTLER 1979 a) which made it necessary for the first
time in the history of air control in NRW to call out a
smog-alarm in the Western Ruhr District.
The average values of sulphate concentrations calculated
from the weekly values show-parallel to the seasonal
differences of pH-values - high winter concentrations and
lower summer values. The frequency distribution of sulphate
concentrations (Fig.4) shows that the median value for the
annual mean is x_{50} = 15 mg SO_4/l, the 10 p.c. value
x_{10} = 9 mg SO_4/l and the 90 p.c. value x_{90} = 26 mg SO_4/l.

The average amount of pollutants, which reaches the soil
per month and hectare, could calculated for calcium
(1.75 kg) and for chloride (2.1 kg).

Considering the pollutant sulphur I'll deal a bit more in
detail with the load per unit area by wet and dry deposition
for the investigated area Central Ruhr District (Fig.5).
The geographical limits of the polluted area with space of
761 km^2 is given by the "Luftreinhalteplan Ruhrgebiet
Mitte".
The total SO_2-pollution in the examinated area makes up
about 312,000 tons per year, of which industry has a share
of 94 p.c., domestic heating and small trade of about 5.6
p.c. and traffic of O.4 p.c. . From the partly estimated
concentrations of pollution it was possible to determine
an average load of SO_2-pollution for all unit areas of one
square kilometer of about O.10 mg SO_2/m^3 in the investigated area for the year 1979.
This average per unit area for the Central Ruhr District
of about O.10 mg SO_2/m^3 can easily be compared with the
annual mean value of 1979, which amounted to O.11 mg SO_2/m^3
at Bochum station. The dry deposition values have been
calculated from this measured mean air concentrations of
sulphur dioxide.

For "it is difficult or impossible to measure the dry
deposition directly, because of the influence of the
specific surface qualities of a collecting funnel or measure-
ing container in the measured values ... measurements of
such an apparatus do not show values for the dry deposition
of a real land surface " (translated from MAYER, 1978,
p. 263).

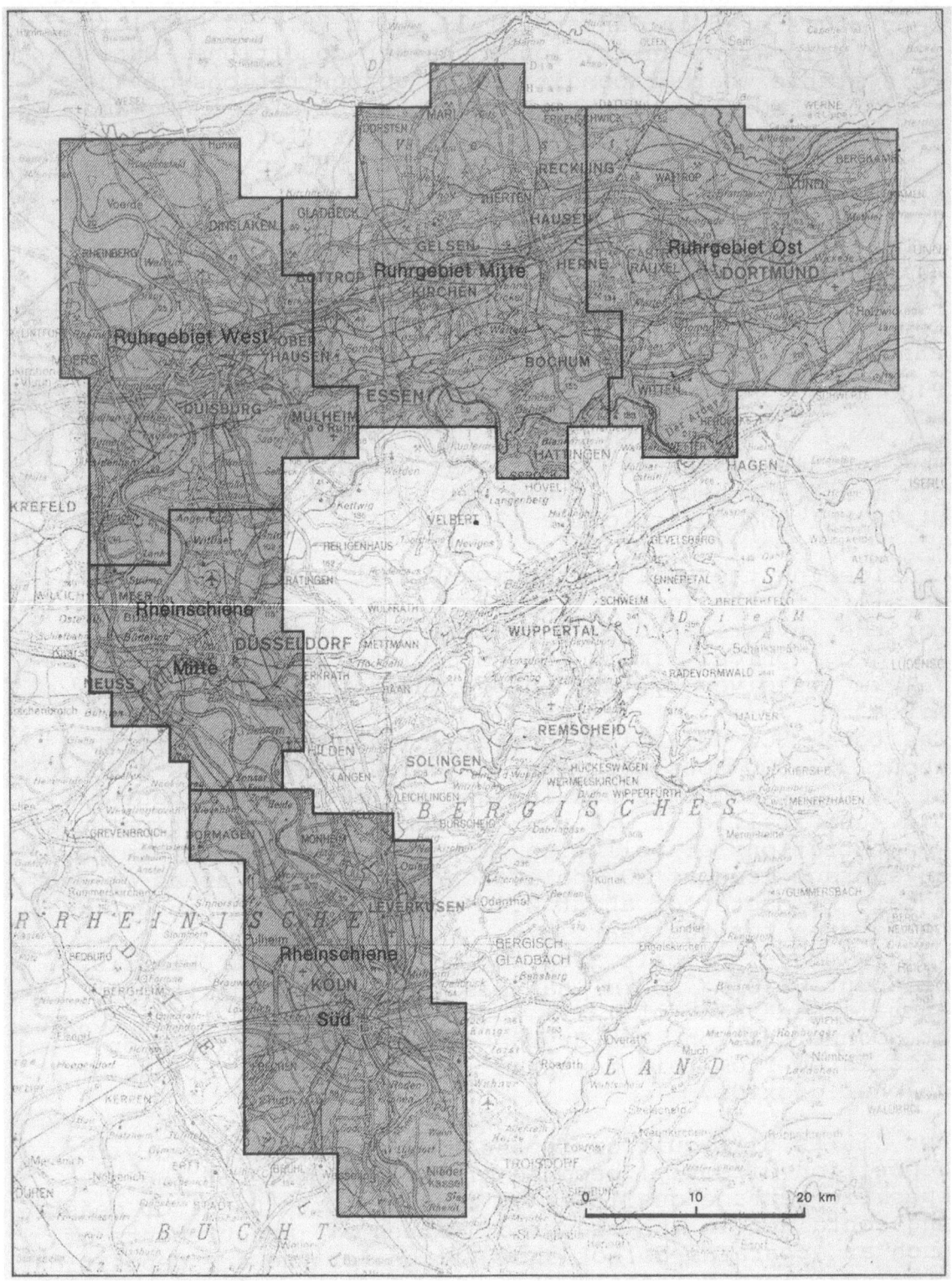

Quelle : Luftreinhalteplan Rheinschiene Süd (Köln) 1977 - 1981.   Grundlage   Ausschnitt aus SK 500 Ü – N –
         Ministerium für Arbeit, Gesundheit und Soziales   Mit Genehmigung des Landesvermessungsamtes NRW
         des Landes NRW (Hrsg.) 1976   vom 12. 10. 1979, Kontrollnummer D 6351
                                                                                      vervielfältigt durch die Ruhr - Universität Bochum.

Fig. 5. Polluted regions in Northrhine-Westfalia
        (KUTTLER 1979b)

The deposition velocity of SO_2 - which have been calculated
for different surfaces - ranged considerably because of
meteorological-, substratum- and plant-specific parameters
(compare the list of PERSEKE and others 1980), so that it
is difficult to come up with the demanded exactness for a
heterogenous structure like the landscape of the investigation area.

Beside artificial and anorganic land surfaces especially
vegetation has an immense influence on the deposition
velocity, this show for example the calculated values by
FOWLER (1980; see Fig. 6).
For closed stomata in vegetation he found a V_g for SO_2
of between o.2 and o.3 cm/s, independent of the height
of the plants. For open stomata, so for proceeding
photosynthesis, he could determine a doubling of deposition
velocity from o.5 to 1.0 cm/s together with an increase of
the height of the plants by factor 100. For wet leaf-surfa-
ces he could even determine a boost of deposition velocity
by the factor 19 with an increase of the height of the
plants by the factor 100.

For it is impossible to determine a corresponding deposition
velocity for every surface and meteorological situation,
we took the generally used value V_g = o.8 cm/s for further
calculations (see PERSEKE and others 1980). In these
estimations it is impossible to consider the seasonal
influence of the vegetation period. The result is shown
in Fig. 7.

The wet deposition of sulphur has been calculated from
the determined sulphate concentrations in the rainwater
for Bochum station in consideration of the precipitation
amount. The result is shown in Fig. 8.

The average monthly wet sulphur deposition amounts to
making up o.29 g S/m^2 with an annual sum of about 3.4 gS/m^2.
In order to be able to express the wet deposition of
pollutants per area in the examined area it was necessary
to know the distribution of the precipitation in this
region. Concerning this matter we made use of the values
of 19 stations for precipitation measurement in the area
of 761 square kilometers. The regional distribution of
precipitation is shown in Fig. 9.
In this map we see a decrease of precipitation from southwest to northwest of about 1000 mm to 750 mm while it
remains nearly unchanged in the direction southeast to
northeast with low values in the Ruhr Valley between 750 mm
and 850 mm.

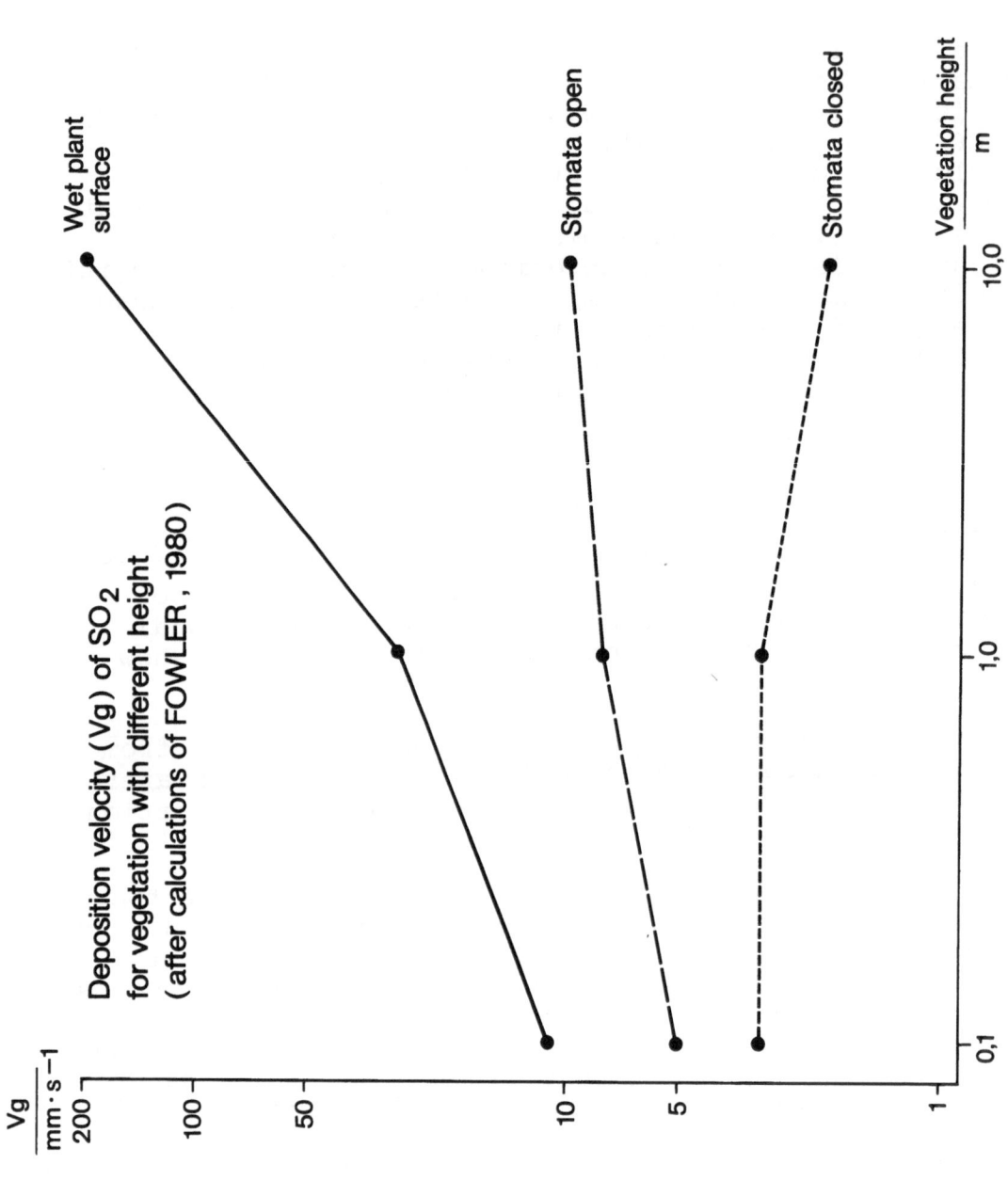

Fig. 6. Depostion velocity of SO_2 for vegetation with different height (after calculations of FOWLER 1980)

Fig. 7   Monthly mean values of sulphur dioxide concentrations and dry SO_2-S Deposition at Bochum station (KUTTLER 1981)

	Jan.	Feb.	März	Apr.	Mai	Juni	Juli	Aug.	Sept.	Okt.	Nov.	Dez.	Jahresmittel bzw. Jahressumme
SO_2 [mg·m^{-3}]	0,26	0,27	0,08	0,07	0,07	0,08	0,06	0,08	0,11	0,11	0,11	0,10	0,11
SO_2-S [mg·m^{-2}]	2785	2612	856	726	750	726	856	642	829	1178	1140	1071	14.171

Fig. 8 SO_4^{2-}-S concentrations in precipitation, monthly precipitation values (N) and monthly wet sulphur depositions (S_{wet}) at Bochum station (KUTTLER 1981)

	Jan.	Feb.	März	Apr.	Mai	Juni	Juli	Aug.	Sept.	Okt.	Nov.	Dez.	Jahressumme bzw. Jahresmittel
SO_4^{2-}-S [mg·l^{-1}]	14,0	8,3	5,3	4,0	4,0	5,7	2,7	2,7	4,0	7,3	4,7	3,3	5,5
N [mm]	26	37	131	47	72	47	63	85	34	16	72	102	732
S_{wet} [mg·m^{-2}]	364	307	694	188	288	268	170	229	136	117	338	337	3436

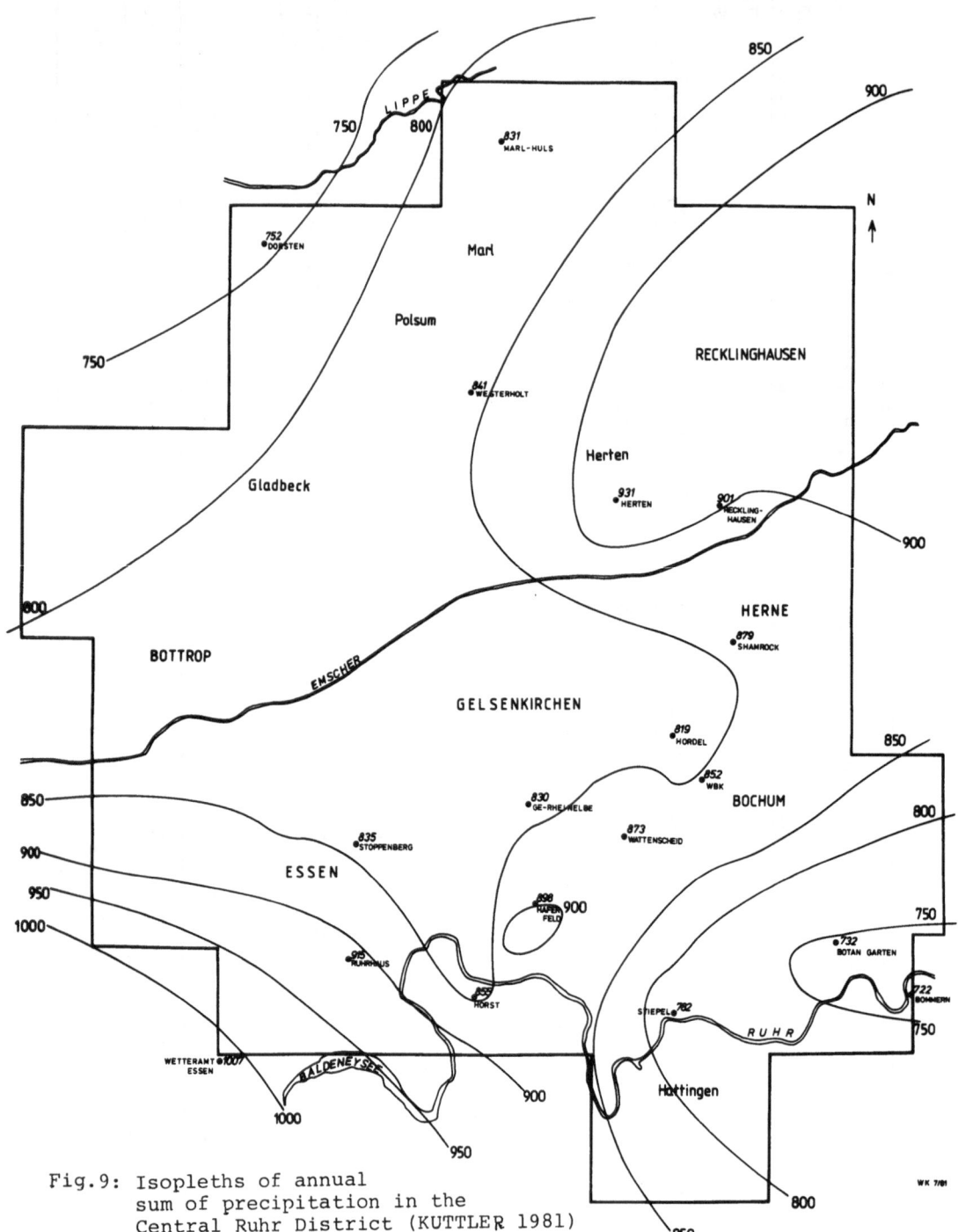

Fig.9: Isopleths of annual
       sum of precipitation in the
       Central Ruhr District (KUTTLER 1981)

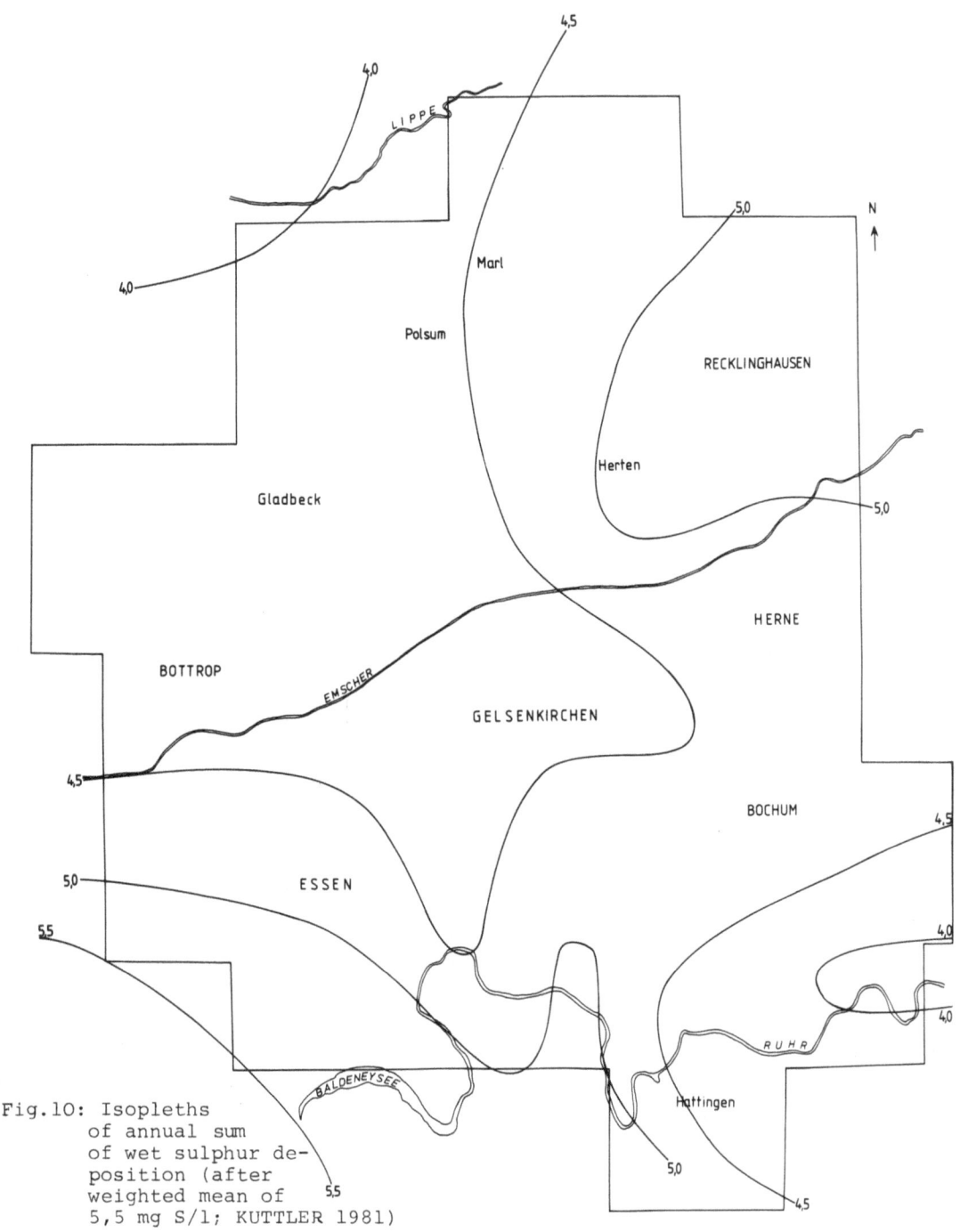

Fig.10: Isopleths
of annual sum
of wet sulphur de-
position (after
weighted mean of
5,5 mg S/l; KUTTLER 1981)

Fig. 11   Dry, wet and total sulphur deposition in 1979 at Bochum station (KUTTLER 1981)

	Jan.	Feb.	März	Apr.	Mai	Juni	Juli	Aug.	Sept.	Okt.	Nov.	Dez.	Jahres-summe
$SO_2 - S_{dry}$ $[mg \cdot m^{-2}]$	2785	2612	856	726	750	726	856	642	829	1178	1140	1071	14.171
$SO_4^{2-} - S_{wet}$ $[mg \cdot m^{-2}]$	364	307	694	188	288	268	170	229	136	117	338	337	3.436
$\sum$ wet, dry $[mg \cdot m^{-2}]$	3149	2919	1550	914	1038	994	1026	871	965	1295	1478	1408	17.607
prozentualer dry	88	89	55	79	72	73	83	74	86	91	77	76	80
Anteil wet	12	11	45	21	28	27	17	26	14	9	23	24	20

From the annual sums of precipitation of the stations and the knowledge of the sulphur concentrations in precipitation we prepared a further map of the annual sum of wet sulphur deposition. This map (Fig. 10) shows that the highest concentrations of pollutants have been deposited in the southwest with values of 5.5 g S/m^2 and year. The lowest depositions of pollutants are found in the northwest and southeast.

The following table (Fig. 11) serves a survey of the dry, wet and total sulphur deposition as well as the dry and wet deposition percentage of the total sulphur deposition in the Ruhr District.

Referring to the amount of SO_2 - S pollution of about 156,000 tons in the Central Ruhr District in 1979, about 9 p.c. of the sulphur is deposited within this region.

We plan to extend the investigations for the whole Ruhr District and its periphery when we'll have gained further results from the particular stations.

3. Literatur

FOWLER, D. (1980): Removal of sulphur and nitrogen compounds from the atmosphere in rain and by dry deposition.- In: Proc. Int. Conf. Impact acid precipitation, Norway, SNSF project, p.22-32.

GEORGII, H.W., PERSEKE, C., ROHBOCK, E. &GRAVENHORST, G. (1980): Untersuchung über die trockene und feuchte Deposition von Luftverunreinigungen in der Bundesrepublik Deutschland. Luftreinhaltung, Forschungsprojekt 10402600. Im Auftrag des Umweltbundesamtes.

GLASS, N. R., POWERS, C.F., LEE, J.J., RAMBO, D.L. & O'GUINN, D.W. (1980): The sensitivity of the United States environment to acid precipitation.- In : Proc. Int. Conf.Impact acid precipitation, Norway, SNSF project, p. 114-115.

GÜNTHER, K.H. & KNABE, W. (1976): Messungen der Schwefel- und Säureniederschläge im Ruhrgebiet in der Zeit von Juli 1973 bis März 1974.- Schriftenreihe der Landesanstalt für Immissions- und Bodennutzungsschutz des Landes Nordrhein-Westfalen, Essen, Heft 39, p. 36-44.

HUTTUNEN, S. (1979): The integrative effects of air-borne
        pollutants on boreal forest ecosystems.- Sym-
        posium on the effects of air-borne pollution
        on vegetation, Warsaw (poland), 20.-24. Aug.
        1979, p. 1-20.

KAYSER, K., JESSEL, U., KÖHLER, A. &RÖNICKE, G. (1974): Die
        pH-Werte des Niederschlages in der Bundesre-
        publik Deutschland 1967 - 1972.- Deutsche
        Forschungsgemeinschaft, Mitteilung IX der Kom-
        mission zur Erforschung der Luftverunreinigung,
        Boppart.

KLOCKOW, D.(1978): Zum Zusammenhang zwischen pH-Wert und
        Elektrolytzusammensetzung von Niederschlägen.-
        VDI-Berichte Nr. 314, p.21-26.

KUTTLER, W. (1979 a): London-Smog und Los Angeles-Smog.-
        In: Erdkunde, Bd. 33, p. 236-240.

KUTTLER, W. (1979 b): Einflußgrößen gesundheitsgefährden-
        der Wetterlagen und deren bioklimatische Aus-
        wirkungen auf potentielle Erholungsgebiete -
        dargestellt am Beispiel des Ruhrgebietes und
        des Sauerlandes. Bochumer Geographische Arbei-
        ten, Band 36, 130 p., Schöningh, Paderborn.

LUFTREINHALTEPLAN RUHRGEBIET MITTE 1980 - 84 (1980):
        Herausgegeben vom Ministerium für Arbeit, Ge-
        sundheit und Soziales des Landes Nordrhein-
        Westfalen.

MAYER,R. (1978): Nasse und trockene Deposition von Schwefel-
        verbindungen auf industrieferne Wälder und
        ihre Wirkung auf den Boden.- VDI-Berichte Nr.
        314, p. 263-265.

MOLDAN, B. (1980): The analysis of atmospheric precipita-
        tion in Czechoslovakia.- In: Proc. Impact
        acid precipiation, Norway, SNSF project,
        p. 124-125.

PERSEKE, C., BEILKE, S. & GEORGII, H.W. (1980): Die Gesamt-
        schwefeldeposition in der Bundesrepublik
        Deutschland auf der Grundlage von Meßdaten des
        Jahres 1974. Berichte des Instituts für Meteo-
        rologie und Geophysik der Univ. Frankfurt/M.

ROOT, J., McCOLL, J. & NIEMANN, B. (1980): Map of areas
        potentially sensitive to wet and dry deposi-
        tion in the United States.- In: Proc. Int.
        Conf. Impact acid precipiation, Norway, SNSF
        project, p. 128-129.

ULRICH, B., MAYER, R., KHANNA, P.K., SEEKAMP, G. &
        FASSBENDER, H.W. (1976): Input, Output und
        interner Umsatz von chemischen Elementen bei
        einem Buchen- und einem Fichtenbestand .-
        Verh. Ges. f. Ökologie, Göttingen, p. 17-28.

KUTTLER, W. (1981): Trockene und nasse Schwefeldepositionen
        im mittleren Ruhrgebiet.- Vortrag auf der 11.
        Jahrestagung der Gesellschaft für Ökologie in
        Mainz, zum Druck eingereicht.

Summary:

In this report the results of rainwater measurements of
different pollutants (sulphur, calcium, chloride and the
pH-value from the station Bochum, Central area of the
Ruhr District, Northrhine-Westfalia) are discussed.
For this area,moreover, calculations about wet and dry
deposition are made for the pollutant sulphur.
The results are presented in several figures and in a map.

NEUTRALIZATION OF ACID IN PRECIPITATION AND SOME RESULTS OF SEQUENTIAL RAIN SAMPLING

Willem A.H. Asman and Piet J. Jonker
Institute for Meteorology and Oceanography
5 Princetonplein, 3584 CC Utrecht, The Netherlands.
Jakob Slanina and Jan H. Baard
Netherlands Energy Research Foundation
P.O. Box 1, 1755 ZG Petten, The Netherlands.

It is shown that the acid content of rain periods associated with maritime air masses does not differ much from rain periods associated with continental air masses. This is mainly due to the much higher degree of neutralization of acid by NH_3 during the latter periods. By using two coupled sequential rain samplers it was shown that the high variability of concentrations during a precipitation event is a real effect, far more important than that caused by statistical variability. Furthermore it is shown that ratios of concentrations of components are not constant within a precipitation event.

NEUTRALIZATION OF ACID IN PRECIPITATION

Dr. P. Winkler from the Deutscher Wetterdienst at Hamburg presented at this colloquium a lecture on the pH-value of rain-water during precipitation events and the origin of the associated air mass. He was not able to find a large difference between the pH-values originating from continental and maritime air masses. This is in accordance with measurements we performed at a coastal site near Den Helder, The Netherlands (Asman, Slanina and Baard, 1981). In this experiment rain-water was sampled with 8 wet-only samplers at one site during periods of 2-3 days and in most cases all bulk-components were determined (H, NH_4, Ca, Mg, K, Na, SO_4, NO_3, Cl). The periods were classified in two groups as originating from continental air masses and as originating from maritime air masses including air which had passed over Great Britain and Ireland. For all bulk-components excluding H and K, a significant difference could be found between the two groups at a significance level of 5% (one-tailed). For H the difference was found at a significance level of 15% (one-tailed).
The results of both experiments seem to be surprising at first sight, as one would expect much higher H-concentrations to occur during continental periods than during maritime periods. This because all main sources of acidifying substances (SO_2, NO_x) are found on land. But the situation is more complicated as not only acidifying substances are found in the atmosphere but also alkaline substances like NH_3 and

115

H.-W. Georgii and J. Pankrath (eds.), Deposition of Atmospheric Pollutants, 115–123.

$CaCO_3$ of which the main sources also are found on land. These alkaline substances are able to neutralize at least part of the acid found in precipitation.

The following relation between concentrations of components can be derived theoretically for precipitation (Asman, Slanina and Baard, 1981)

$$[H] + [NH_4] + 2[Ca^*] + 2[Mg^*] + [K^*] = 2[SO_4^*] + [NO_3] \qquad (1)$$

(all concentrations in mole.1^{-1}; $*$ = corrected for sea spray)

In our experiment the difference between the theoretical relation and the computed regression function was very small (Asman, Slanina and Baard, 1981). We now assume, just as Granat (1972) did that for each mole of NO_3 one mole of H was originally available, for each mole of SO_4^* two moles of H and that NH_4, Ca^*, Mg^* and K^* are the results of neutralizing 1, 2, 2 and 1 moles of H respectively. Then it is possible to compute the relative importance of neutralizing substances.

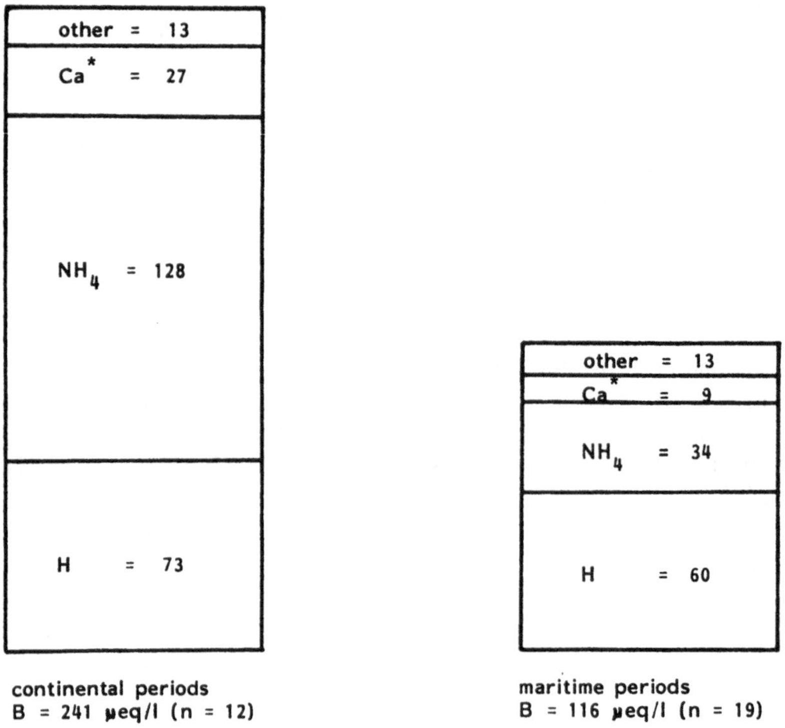

other	=	13
Ca*	=	27
NH$_4$	=	128
H	=	73

continental periods
B = 241 μeq/l (n = 12)

other	=	13
Ca*	=	9
NH$_4$	=	34
H	=	60

maritime periods
B = 116 μeq/l (n = 19)

Figure 1. Weighted average concentrations of some cations for continental and maritime periods.

Figure 1 shows the weighted average concentrations (in μeq.1^{-1}) of the cations of equation (1) for continental and maritime periods. The total height of the bars represents the sum of SO_4^* and $NO_3 = B$ (in μeq.1^{-1}). This value was taken instead of the sum of H, NH_4, Ca^*, Mg^* and K, because not all Mg^* and K^* values were determined. But this does not matter much since both sums seem to be equal. The H-concentration does not differ very much for both periods, but B and the NH_4-concentration do.

The relative importance of NH_3 as a neutralizing agent for H-ions can be expressed as a percentage:[3] $100 \cdot [NH_4]/B$. For continental periods the relative neuralization of acid by NH_3 amounts to 53% and for maritime periods to 29%. This implies that although the amount of acid generated by SO_2 and NO_x is much larger for continental periods than for maritime ones, the amount of acid after neutralization is almost equal because during continental periods a higher degree of neuralization is observed. Calcium-and other components seem to be less important as neutralizing agents than NH_3. More recent experiments of our project confirm these observations.

One has, however, to be careful with conclusions with respect to the effects of the precipitation. There are strong indications that in soil NH_4 can be oxidized to HNO_3 (Wiklander, 1980 ; Rosenqvist, Jørgensen and Rueslätten, 1980). The overall reaction is:

$$NH_4^+ + 2 \ O_2 \rightarrow 2 \ H^+ + NO_3^- + H_2O$$

If this process takes place there will be a large difference in impact of precipitation associated with continental and maritime periods with respect to acidification.

STATISTICAL VARIABILITY DURING SEQUENTIAL SAMPLING

Galloway and Likens (1978) pointed out that three sources are responsible for the statistical variability in data in precipitation chemistry: analytical, sampling and spatial sources. In addition storage can also be a source of statistical variability because concentrations can alter if samples are stored incorrectly or for a long time.

To determine the statistical variability when collecting precipitation with wet-only samplers at one site we performed experiments with 8 of such samplers at one site. With a sequential rain sampler rain-water is collected sequentially, usually from one precipitation event only. Concentrations during single precipitation events are highly variable (Asman, 1980). Therefore it was sometimes suggested that this phenomenon was caused to a large extent by the statistical variability and not by any real effect. To examine whether this was the case or not and to determine the statistical variability at one site it would be preferable to do an experiment with 8 or more sequential rain samplers. As sequential rain samplers are very expensive this was not possible. To get at least an indication of the possible variability we did an experiment with two sequential rain samplers.

Each sequential rain sampler consisted of a polyethylene funnel with an opening of $0.19 \ m^2$. The funnel was covered by a pneumatically operated lid activated by a rain detector. This technique prevents dry deposition in the sampler. The rim of the funnel was placed 1.50 m above the ground. The funnel and the lid were cleaned at least once a

day. The funnel was connected by a polyethylene tube with a fraction
collector with sample tubes. A sample tube was removed when a volume
of 20 ml was reached, corresponding to a rainfall depth of about 0.1 mm.
A digital printer was used to record the beginning and the end of rain
events, and the time at which sample tubes were removed. This enables
the computation of the rainfall rate.
If samples collected with two sequential rain samplers have to be
compared, they have to be taken during the same period. As the
rainfall rate also shows a variability on a small spatial scale, this
implies that samples (with the same volume) obtained with two samplers
are generally not taken during the same period. This variability also
causes that the lids of the two samplers are not operated at the same
time and hence that the samples cannot be compared.
To overcome these problems only one rain detector was used, connected
to both samplers, so that the lids were removed at the same time.
Moreover, one of the sequential samplers functioned as a master sampler:
if a tube of this sampler was changed the tube of the other one was
changed automatically at the same time. In that way we were able to
obtain couples of comparable samples. The distance between the two
samplers was 4 m during the experiment.
Figures 2-6 show the first results of the experiment. Apart from a few
outliers, the concentrations of different components collected with
the samplers did not differ much.

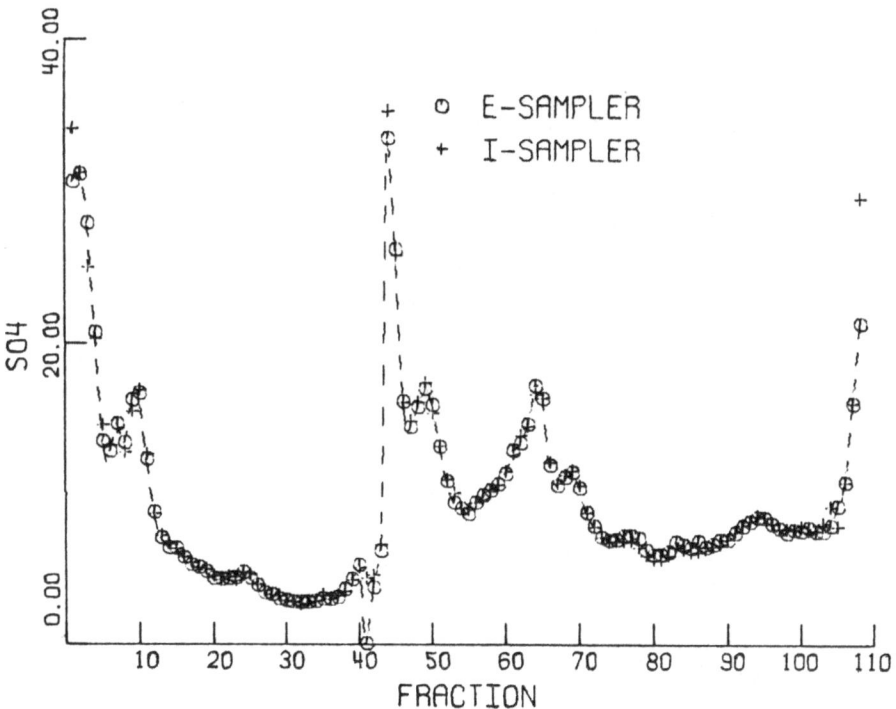

Figure 2. SO_4-concentration (ppm).

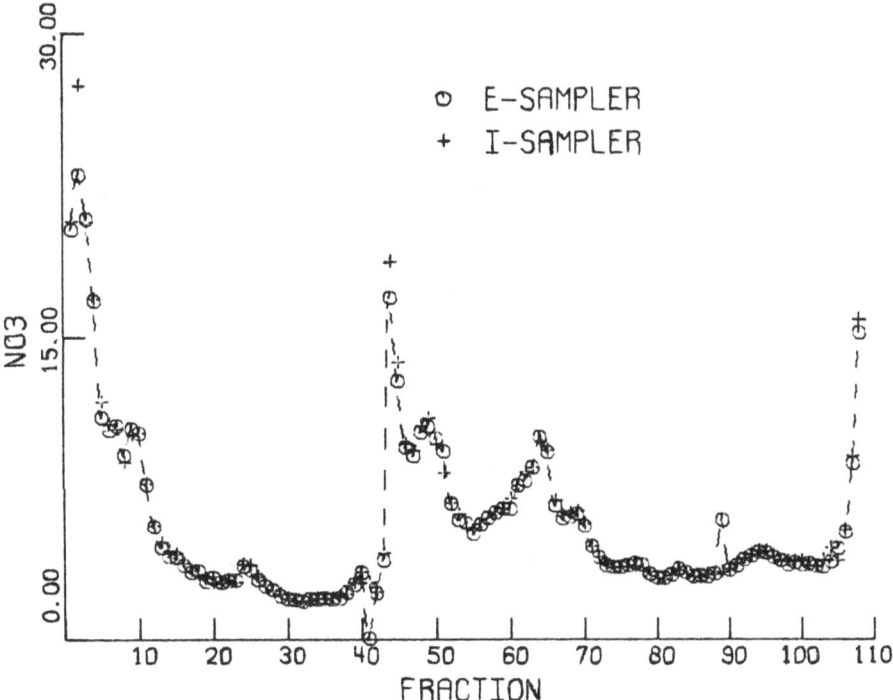

Figure 3. NO$_3$-concentration (ppm).

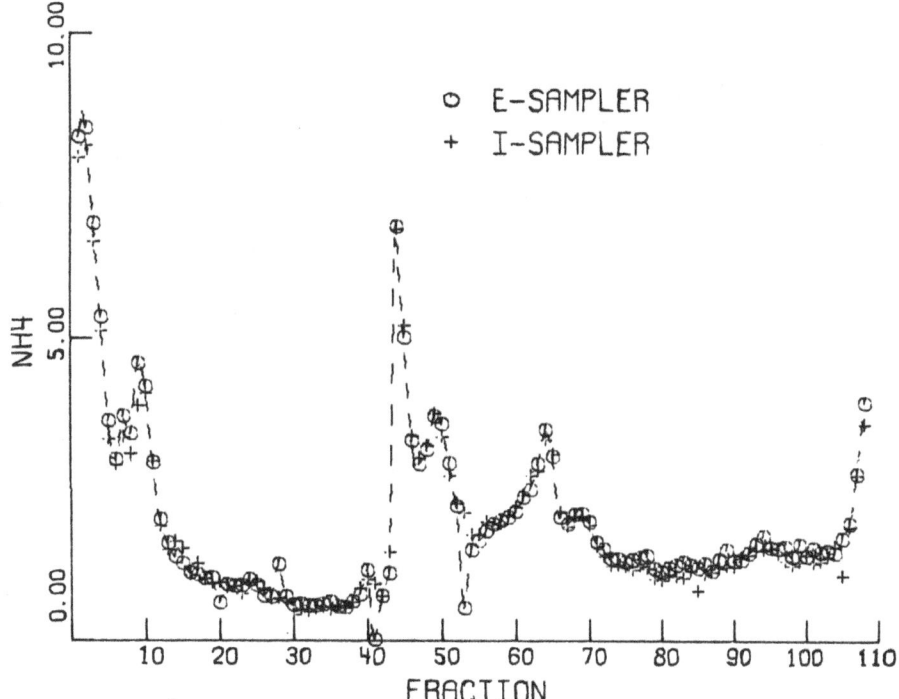

Figure 4. NH$_4$-concentration (ppm).

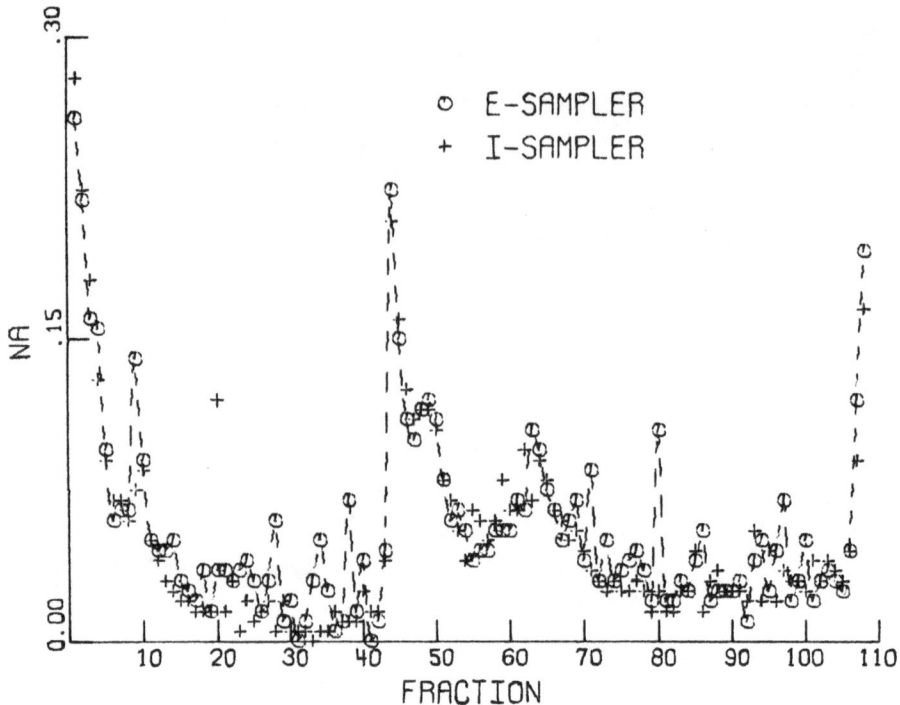

Figure 5. Na-concentration (ppm).

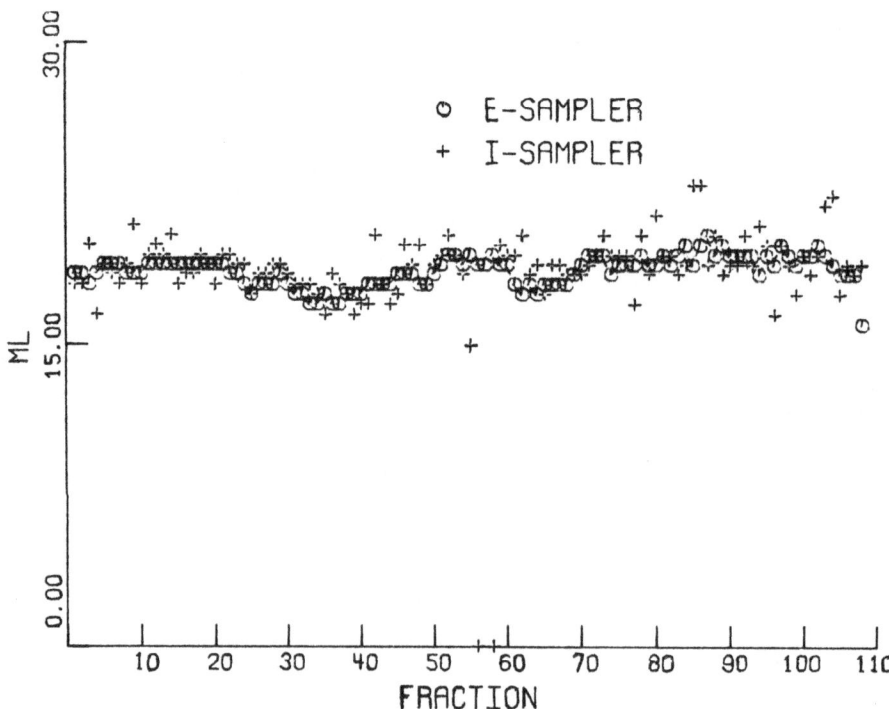

Figure 6. Sample volume (ml).

It should be kept in mind that the volume of the samples collected
with the master sampler (denoted by circles) was kept constant and
hence the volume of the other sampler would be more variable. Minor
changes in the volume of the samples collected with the master sampler
occur during high rainfall rate because the tubes are then filled
noticeable during the change. No statistical evaluation was yet made.
These preliminary results indicate that the statistical variability is
not large and that highly variable concentrations occur in reality.
Therefore it is possible to start with the (meteorological) interpretation of the results of sequential rain sampling.

INTERPRETATION OF THE RESULTS OF SEQUENTIAL RAIN SAMPLING

Sequential rain sampling is applied to study scavenging processes in
the atmosphere. It was generally thought that the changes in
concentration during precipitation events could give information on
these scavenging processes. But the results until now seem to be
disappointing, mainly because concentrations in precipitation are the
result of the influence of many factors acting at the same time.
Each precipitation event seems to have its own character and is
difficult to compare with other events. Another possibility is then
not to try to draw conclusions from the changes in concentration for
one component, but to compare concentrations of different components
belonging to the same event. This, for example, is done by correlating
concentrations of components (Dawson, 1978).
One has to be careful if high correlations are found. It is tempting
then to conclude that the ratio of the concentration of the two
components is constant. This is generally not true. If the regression
function passes the origin, the ratio of the two components is constant,
but if the regression function does not, the concentration ratio c_2/c_1
of two components will be a function of time (or fraction number).
In this case the concentration of component 2 is: $c_2 = a.c_1 + b$
(a and b are constants). If the concentration of component 1 is a
function of time, the ratio c_2/c_1 will also be a function of time and
is hence not constant: $c_2/c_1 = a + b/f(t)$. This is illustrated by
Figure 7.

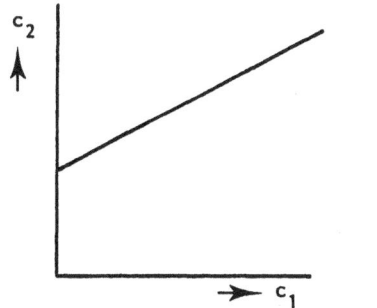 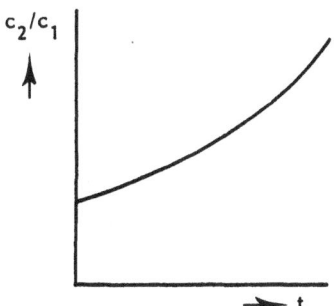

Figure 7. Regression of c_2 on c_1; time dependence of c_2/c_1.

In our experiment we tried first to compare concentrations of two
components by plotting them on a logarithmic scale versus the sample
number. If the shape of both curves is exactly the same, their
concentration ratio should be constant.
Even for sea salt components like Na and Cl this was not the case,
because there is always some statistical variability in the data.
Therefore we developed a procedure by which concentration curves are
compared and statistical variability is taken into account. In this
procedure the concentration of one component of each sample was
normalized by dividing it by the mean value of the component for the
whole event: $c_1/c_{1\ mean}$. This was also done for the other component.
Then the ratio $(c_1/c_{1\ mean})/(c_2/c_{2\ mean})$ was computed for each sample.
Only if for 3 or more samples this ratio was significantly different
from the average ratio $c_{1\ mean}/c_{2\ mean}$ it was plotted versus the
sample number for all samples of the event.
In fact not the arithmetic averages $c_{1\ mean}$ and $c_{2\ mean}$ were computed,
but weighted averages, taking into account the error in each concentration value. Preliminary results of this procedure for a shower
associated with a trough are presented in Figures 8 and 9.
The vertical bars indicate the error in the ratio. The results show
that many concentration ratios are not constant during this shower.
It is hoped that by combining these data with meteorological data
more information can be obtained on the scavenging processes.

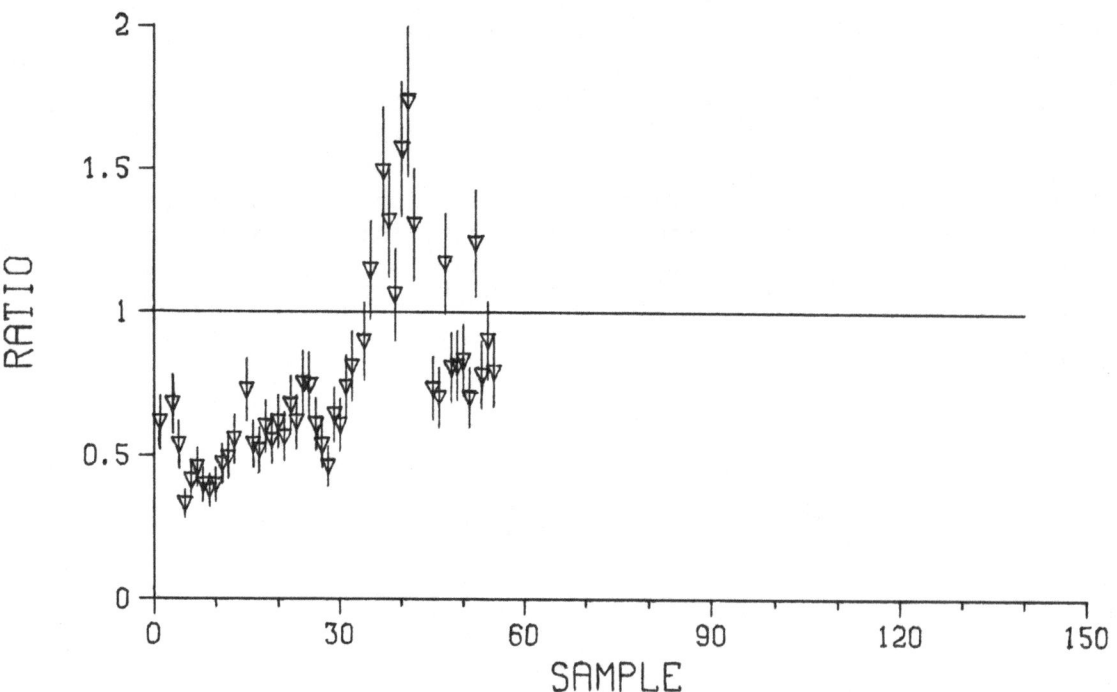

Figure 8. Normalized SO_4^*/NO_3-ratio vs. sample number.

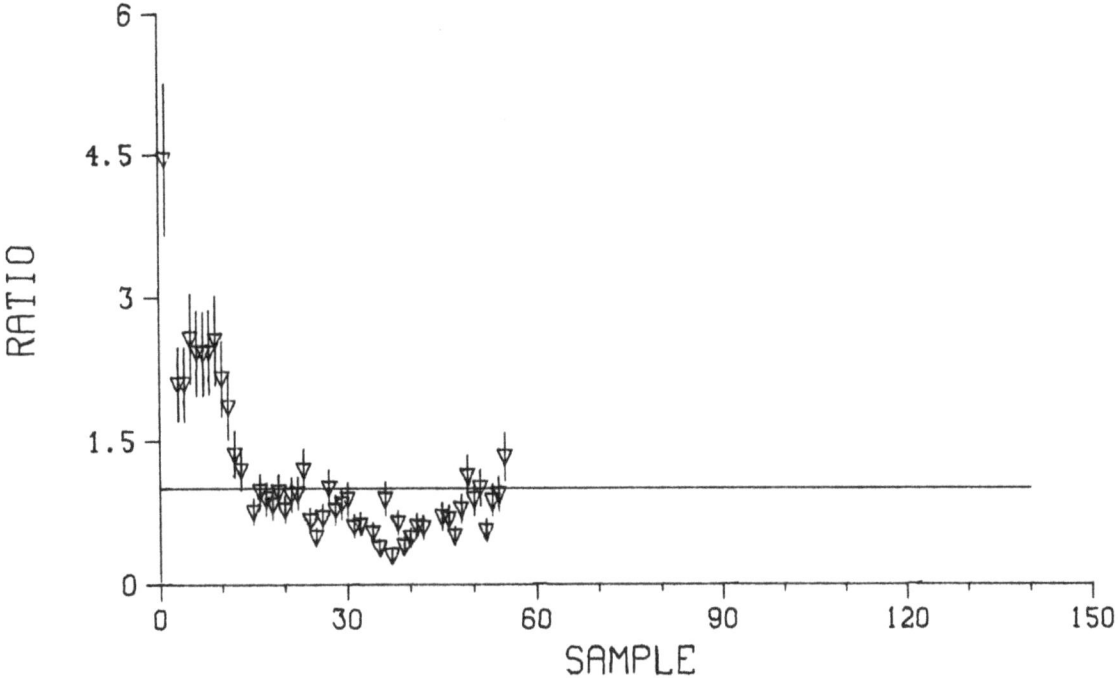

Figure 9. Normalized Na/SO_4^*-ratio vs. sample number.

ACKNOWLEDGEMENT

This research is part of a joint project on precipitation chemistry, being carried out by the Netherlands Energy Research Foundation and the Institute for Meteorology and Oceanography of the State University of Utrecht.
The authors wish to express their thanks to Mr. Ed Buijsman and Mrs. Iris E. Odlum for their help by taking some sequential rain samples. One of us (Willem A.H. Asman) received financial assistance from the Ministry of Health and Environmental Protection.

REFERENCES

Asman, W.A.H.: 1980, Water, Air and Soil Poll. 13, pp. 235-245.
Asman, W.A.H., Slanina, J. and Baard, J.H.: 1981, Water, Air and Soil Poll. 16, pp. 159-175.
Dawson, G.A.: 1978, Atmospheric Environment 12, pp. 1991-1999.
Galloway, J.N. and Likens, G.E.:1978, Tellus 30, pp. 71-82.
Granat, L.: 1972, Tellus 24, pp. 550-560.
Rosenqvist, I.Th., Jørgensen, P. and Rueslåtten, H. in: Drabløs, D. and Tollan, A. (eds.): 1980, Ecological impact of acid precipitation, Proceedings of an international conference, Sandefjord, Norway, March 11-14, 1980, SNSF-project, Oslo-Ås, pp. 240-241.
Wiklander, L. in: Hutchinson, T.C. and Havas, M. (eds.): 1980,Effects of acid precipitation on terrestrial ecosystems, Plenum Press, New York, pp. 553-567.

THE INFLUENCE OF IMMEDIATE FREEZING ON THE CHEMICAL COMPOSITION OF RAIN-SAMPLES

K.P. Müller and G. Aheimer

Institut for Atmospheric Chemistry
Nuclear Research Centre Jülich

G. Gravenhorst
CNRS Grenoble
Laboratoire de Glaciologie

ABSTRACT

An evaluation of the conversion of some atmospheric trace constituents in airborne droplets or the transport of trace substances by rain to the ground needs the knowledge of the chemical composition of the rainwater. Precipitation - networks generally have to work with time intervals of days up to weeks between sampling and chemical analysis of rainwater[1]. During those periods the original composition can be drastically changed. Variations may be caused by 1: various effects on the walls of sampling containers, 2: chemical reactions and 3: biological conversion in the liquid sample[2]. A precipitation - sampler was modified in that way, that with the starting of the rain-event the sampling container was automatically cooled to $-20^\circ C$ by means of a refrigerator. Simultaneously non cooled precipiation of the same rain was collected for a comparison. By ion-chromatography the soluble ions Cl^-, NO_3^-, and SO_4^- were analyzed. Ammonium was analyzed by photometry[3,4]. It is shown that under normal storage conditions the ammonium concentration strongly varies with storage time. The discrepancy of concentrations of samples analyzed immediately after rainfall and those analyzed after a certain period of storage time can be avoided by immediate freezing of the rainwater during sampling.

1. INTRODUCTION

Precipitation and dry-deposition samples were analyzed at the institute of atmospheric chemistry of the Nuclear Research Centre at Jülich as a part of a project of the Federal Environmental Agency of Germany (UBA) in order to estimate the wet and dry deposition of several trace constituents[5]. In the laboratory at Jülich investigations of

H.-W. Georgii and J. Pankrath (eds.), Deposition of Atmospheric Pollutants, 125–132.

samples were made which were collected at ten measuring
stations located at different places of the Federal Republic
of Germany[5]. The ions we were looking for have been fluoride,
chloride, nitrite, phosphate, bromide, nitrate, sulfate and
ammonium. The nitrite, bromide and phosphate concentration
in rain and aerosols very seldom exceeds the detection limit
(10 ng/ml). The other ions are in the range of nanograms to
mikrograms per milliliter. In view of these low concentrations contamination by improper handling, by the used
containers or by ambient laboratory-air must be thoroughly
avoided. Lowering the detection limits contamination more
seriously becomes a growing problem. Due to this fact contamination is well investigated[3]. Working with clean-air-
benches does not exclude all interfering materials. Only
aerosols are totally retained but gases like NH_3, HCl and
HNO_3 are able to penetrate the filtering materials and
originate considerable contamination. Sequential blank tests
are checks to control the contamination - free sample -
handling.
Not only the mentioned contaminations can disturb trace
analysis but also a possible chemical loss of the trace
substances may seriously affect the investigations[6]. While
detecting ammonium in rainwater we noticed a remarkable
decrease of the concentration with increasing storage time.
The observed reduction can be initiated by interaction of
the liquid sample with the container walls, by radiation
(sunlight) or by temperature influence.

II. POSSIBLE EFFECTS DUE TO STORAGE

To suppress wall-reactions that can take place at the active
points of the wall surface by addition or catalytic
reactions, we first have looked for several container
materials. We have tested vessels made out of polyethylene
(PE), polypropylene (PP), glass and polytetrafluorethylene
(PTEE) storing rain samples with known concentrations over
a period of two weeks[7]. The results are shown in table 1.

Loss of concentration within 14 days (in percent)	
polypropylene	8
polyehtylene	12
glass	51

table 1

PTEE is omitted in the table because of its unreproducible
effects on the ion-concentration and even of its high costs.
Another parameter that influences that stability of rain
samples is the ambient temperature[8]. We have performed
storage spot checks at roomtemperature (RT), at 0° to 4°C

in a refrigerator (RF) and at -18° to $-20^{\circ}C$ in a deep
freezer (DF). We saw first hints that storing the rain
samples at temperatures below the freezing point of water
results in positive effects on the conservation of the
original composition of precipitation (table 2).

Loss of concentration within 14 days (in percent)	
deep-freezer	6
refrigerator	18
room-temperature	78

table 2

The influence of light also was discussed[9], but especially
in the case of ammonium we did not find a significant improvement by dark storage compared with that under normal
conditions. Lighting also causes biological activity[19].
Especially nitrogen compounds like nitrate and ammonium are
decomposible by bacteria. A possible prevention is the
poisoning of the samples by fomaldehyde, chloroform or
mercury-salts. Furthermore acidifieing to about a pH of two
is likewise able to stop bacteria activity.
The best procedure to avoid changes in a positive or negative
direction is the immediate in-situ-analysis of all compounds
of interest. But within precipitation networks with a lot
of collecting sites immediate measurements cannot be performed: for logistic and financial reasons it would be
impossible to implement a network equipped with all required
analyzing systems. Furhtermore prevention of rain samples
by adding poisoning agents cannot be accepted, if most
comprehensive results are desired. They disturb the intrinsic
analysis by simulating additional ion-concentrations in ion-
chromatography or by inhibition of color reactions in photometry. More over the determination of the pH-value in the
rainwater is impossible after adding mineral acids. In
summary there must be any kind of storage.

III. DETAILED INVESTIGATIONS OF TIME DEPENDENT CONCEN-
      TRATION LOSSES
To exclude influences of storage, we first tried to analyze
the most delicate ammonium as soon as possible. The annual
variation shown in figure 3 is consistent with ammonium
concentrations in rainwater that have been analyzed immediately after collection.
The result is only partially representative because only
day time samples or shortly before day break falling rain
could be handled and analyzed. Although an evident increase
of ammonium concentrations in precipitation during summer
months is obvious. In order to study the changing ammonium
concentrations during storage more detailed, we stored
precipitation water obtained during a summer - (high values)

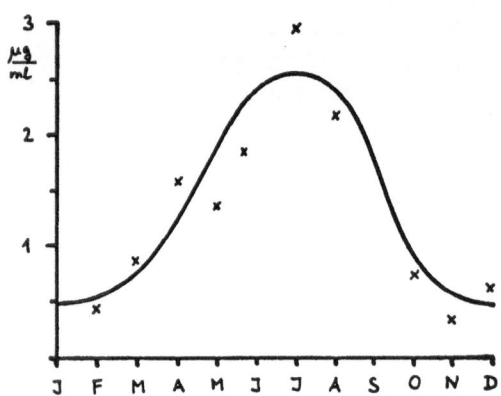

Figure 3

and a winter (low values) rain event in polyethylenflasks
which were shown to have only a little influence on the
composition of rain samles (s. tab. 1).
The ammonium concentration dramatically changed regardless
of storage of samples in darkness or daylight and at
temperatures above $0^{\circ}C$ (fig. 4). The concentration decreased
a time constant of some hours up to two days down to 10
percent. The mean reduction amounted to 50 percent of the
absolute starting concentration.

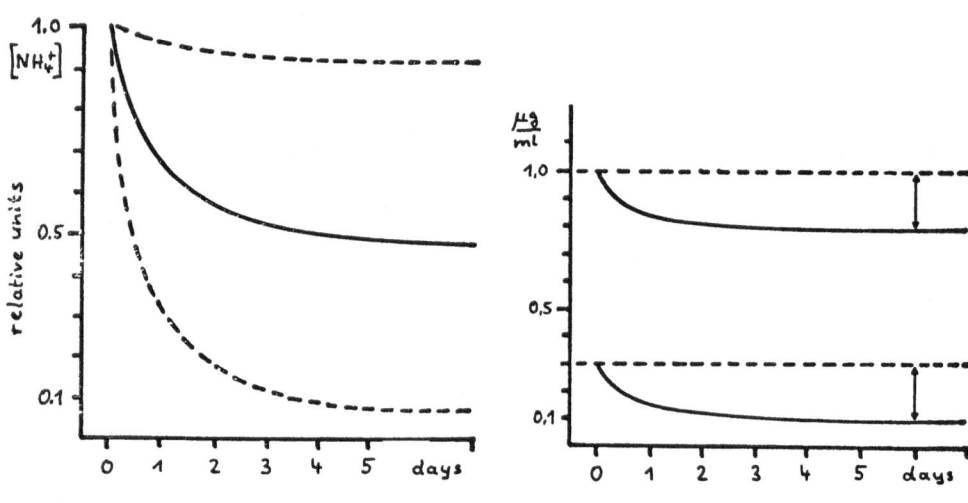

Figure 4   Figure 5

The lower values obtained during winter time show a more significant scattering in the percentural reduction (fig. 5). They have possible losses up to a hundred percent. The essential reason for this effect is the equality of the absolute decrease of concentrations in winter. In the case of lower values the ammonium contents can decline within one day under the detection limit. This diminution by a constant amount is independent from the initial concentration and indicates wall reactions, that end in a saturation state. In Winter less nitrophile bacteria are present than during summer time, when additional biological activity occures. Furthermore subsequent acidifieing causes a more inferior desorption from the vessel-walls than in winter time. Bacteria (for example bacterium pseudomonas) are able to use the ammonium-ion for their metabolism. On the other hand an increase in nitrate has never been noticed:

the ammonium is irreversibly decomposed and lost for the analysis. Already after one or two days of storage of the rainwater samples, which no longer interact with their natural environment, ammonium concentrations decrease by a factor of two. Due to that short lifetime and due to the observed wide range of losses of concentration, storage of rainwater samples is impossible.

IV. IMMEDIATE FREEZING OF RAIN SAMPLES

The immediate freezing of precipitation is a possible solution of these problems. We have developed an automatic deep freezer that freezes the precipitation beginning with the first falling drops. Therefore we combined a refrigerating unit type "Danfoss PW 3 x 7 T2", having a small power consumption of 200 VA, with a deposition-collector of the Frankfurt UBA-project which is described in full detail elsewhere[5]. The refrigerator starts to run simultaneously with the opening of the collecting funnel initiated by a humidity-sensor. The evaporator of the aggregate is formed like a cooling-coil around the sampling vessel. It is insulated to prevent strong icing (fig. 6).
After the automatic start the apparatus can only switched off manually. That occurs while daily changing of the sampling flask.
The immediate deep-freezing has several ovious advantages. Already the first drops of precipitation are present in solid phase and they remain in that state by storage in a commercial freezer until the analysis takes place. This procedure results in a significantly reduced diffusion of the wanted ions to the walls of the vessels: accumulations to the wall will be clearly suppressed. Also chemical reactions and catalytical additions will be of in inferior importance because of the lower reaction-enthalpy and the

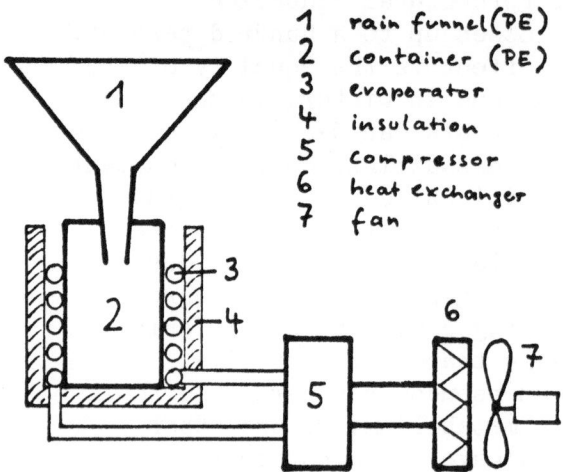

1   rain funnel(PE)
2   container (PE)
3   evaporator
4   insulation
5   compressor
6   heat exchanger
7   fan

Figure 6

smaller probability of hitting in a low-temperature-system.
Furthermore bacteriological activity is surely moderate by
low temperatures and there by the biological decomposition
of nitrogen-containing constituents is dimineshed.
The derivative trends with respect to time of samples stored
at roomtemperature and those stored a $-20^{\circ}C$ are very
different (fig. 7)

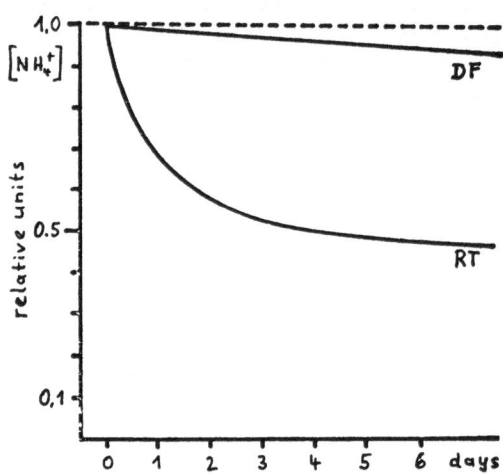

Figure 7

Mainly in summertime the concentrations of precipitation
collected at the same site at the same time already differ
after some hours conspiciously. The unprotected sample
adjusts to the 50 percent level of the starting concentration, while the deep-frozen one decreases only for about
6 percent within 7 days. That loss is supposed to be a
result of the repeated short melting to the purpose of
measurements.

Further experiments will verify this by measuring samples in
parallel. Over a period of 4 weeks no more drastical change
was noticed with respect to both kinds of storage. In this
context we would mentione that we have not performed
investigations of samples beeing exposed for 4 weeks open
to the ambient air, like it commonly is done for monthly
averages[11].

V.   CONCLUSIONS

The immediate freezing offers a contamination - free
preservation of aqueous samples without additional chemical
change by adding auxiliary means.

Besides the described investigations regarding the ammonium-
ion we found that other ions like chloride, nitrate and
sulfate also show a diminution of concentration during non
frozen storage. The effect is clearly visible at lower concentration levels, whereas at higher concentrations the
loss is barely noticeable because of the smaller percentual
decrease.

The clearest change in concentration always shows the
ammonium-ion. Our investigations led us to the conclusion
that the absolute values of the ammonium in rainwater
measured up to now may be too low at least by a factor of
two. Furthermore this will result in the suggestion that
the ammonium would be more important as a counterion to the
sulfate in the ion-balance of precipitation.

ACKNOWLEDGEMENT

This investigation was partially supported by the Federal
Environmental Agency of Germany (UBA), Project-No. 104 02600.

REFERENCES

[1]Granat, L.,: 1974, On the variability of rainwater compo-
    sition and errors in estimates of areal wet deposition,
    USAEC-Conf. 74 1003
[2]Heron, J.,: 1962, Determination of phosphate in water after
    storage in polyethylen, Limnol. Oceanogr., 7, pp. 316-321
[3]Grasshoff, K.,: 1976, Methods of Seawater Analysis, Verlag
    Chemie

[4]Grasshoff, K., Koroleff, F.,: 1977
   Report of the Baltic Intercalibration Workshop Kid,
   7.-19. March 1977

[5]Georgii, H.W., Gravenhorst, G., Perseke, C., Rohbock, E.,
   Untersuchung über die trockene und feuchte Deposition
   von Luftverunreinigungen in der Bundesrepublik Deutschland
   Universitäts-Institut für Meteorologie und Geophysik,
   Frankfurt/Main, UBA-Forschungsprojekt 104 02600,
   Oktober 1980

[6]Dawson, G.A.,: 1978, Ionic composition of rain during
   sexteen convective showers, Atmos. Envir. 12,
   pp. 1991-1999

[7]Furch, K.,: 1975, Die Stabilität wäßriger Ammonium- und
   Phosphat-Lösungen bei der Aufbewahrung in Polyäthylen-
   Gefäßen

[8]Smith, J.P., Grosjean, D., Pitts, J.N,: 1978
   Observation of Significant Lo-ses of Particulate Nitrate
   and Ammonium from High Volume Glass Fiber Filter Samples
   Stored at Room Temperaturen, J. of the Air Pollution
   Control Association 28, pp. 930-933

[9]Meetnet voor Bepaling van de chemische Samenstelling van de
   Neerslag in Nederland, Jaaroverzicht 1980

[10]Degobbis, D.,: 1973, On the storage of seawater samples for
   ammonia determination, Limnol. Oceanogr. 18, pp. 146-150

[11]Galloway, J.N., Likens, G.E.,: 1978, The collection of
   precipitation for chemical analysis, Tellus 30, pp. 71-82

SEASONAL AND REGIONAL DISTRIBUTION OF POLYCYCLIC AROMATIC HYDROCARBONS IN PRECIPITATION IN THE RHEIN-MAIN-AREA

Günther Schmitt
Institute of Meteorologie und Geophysics,
University of Frankfurt/Main

ABSTRACT

Measurements of polycyclic aromatic hydrocarbons (PAH) in precipitation by high-performance liquid chromatography (HPLC) show that precipitation is the dominant sink in the atmospheric PAH-cycle. In the precipitation the PAH-components show a chracteristical distribution. The range of concentration is directly influenced by the meteorological parameters. Significant maximum concentrations in winter and low concentrations in summer prove seasonal variation. The regional distribution of the PAH in precipitation is governed by the influence of local sources.

INTRODUCTION

Polycyclic aromatic hydrocarbons (PAH) are formed in pyrolytic combustion processes of fossil fuels. The main sources are industrial processes, domestic heating and automobile traffic. Combustion material and combustion temperature determines the PAH profile and absolute amounts. In the atmosphere, they are adsorbed on air-borne particulate matter with an aerodynamical diameter ranging from 0,5 to 3,0 µm (1). Besides of nitrosamines, aflatoxines and some heavy metals PAH are potential cancerogenic substances in our environment (2). The main sinks in the atmospheric PAH-cycle are photooxidation and wet deposition by precipitation.

Up to now there exist only few studies on the PAH-concentration in precipitation and their behavior in an aqueous system. Aim of the study was it to present an overview on the PAH in precipitation and to characterize the influence of the meteorological parameters.

H.-W. Georgii and J. Pankrath (eds.), Deposition of Atmospheric Pollutants, 133–142.
Copyright © 1982 by D. Reidel Publishing Company.

The PAH were measured by high-performance liquid chromatography (HPLC) and fluorescence detection. A special
method of enrichment of the soluble PAH-fraction of the
precipitation was necessary.

EXPERIMENTAL

After sampling with a pre-purified funnel (0,25 m^2) the
precipitation is filtrated to separate the insoluble fraction. The insoluble fraction is treated with 75 ml Tetra-
hydrofuran and reduced to exactly 0,5 ml in volume by means
of a rotary-evaporator (3). From the soluble fraction
500 ml were enriched by a liquid-extraction with 3 x 30 ml
cyclohexane. This solution is also concentrated to 0,5 ml.

Both fractions are analysed by HPLC with reverse-phase-sys-
tem RP 8(7 μm particle size). A methanol-water-mixture is
used as mobile phase for gradient elution. The separated
PAH-components are detected with a fluorescence-detector
operating at an extinction-wavelength of 322 nm and an emis-
sion-wavelength of > 350 nm.
Qualitative identification of the PAH compounds is achieved
by comparison of their characteristic retention times with
authentic standards. PAH-concentrations are calculated by
comparing peak areas of calibration curves obtained by use
of standard solution at different concentrations.
A typical chromatogramm of the soluble fraction in precipitation show the separated PAH (Fig. 1).

RESULTS AND DISCUSSION

The distribution of the PAH concentration in precipitation is shown by the monthly mean value of February 1981
(Fig. 2). This diagram represents the compounds in se-
quence of elution as the sum of soluble and insoluble fraction. The concentrations of the single components range
from only few ng/l to some hundreds ng/l. The compound
fluoranthene shows the highest concentration with more than
630 ng/l. The lowest concentrations are analysed for coronene with 35 ng/l. The total amount of the measured PAH
contains 1,8 μg/l. The concentrations are directly influenced by the meteorological parameters. Concentrations
deviate more than 40% from the monthly mean values. Increased values are found at low temperatures and after long
dry periods decreased concentrations at high temperatures
and short dry intervals.
For direct comparison, the PAH-profiles are represented in
a percentage distribution instead of the absolute concentration values (Fig. 3). The different distribution of the

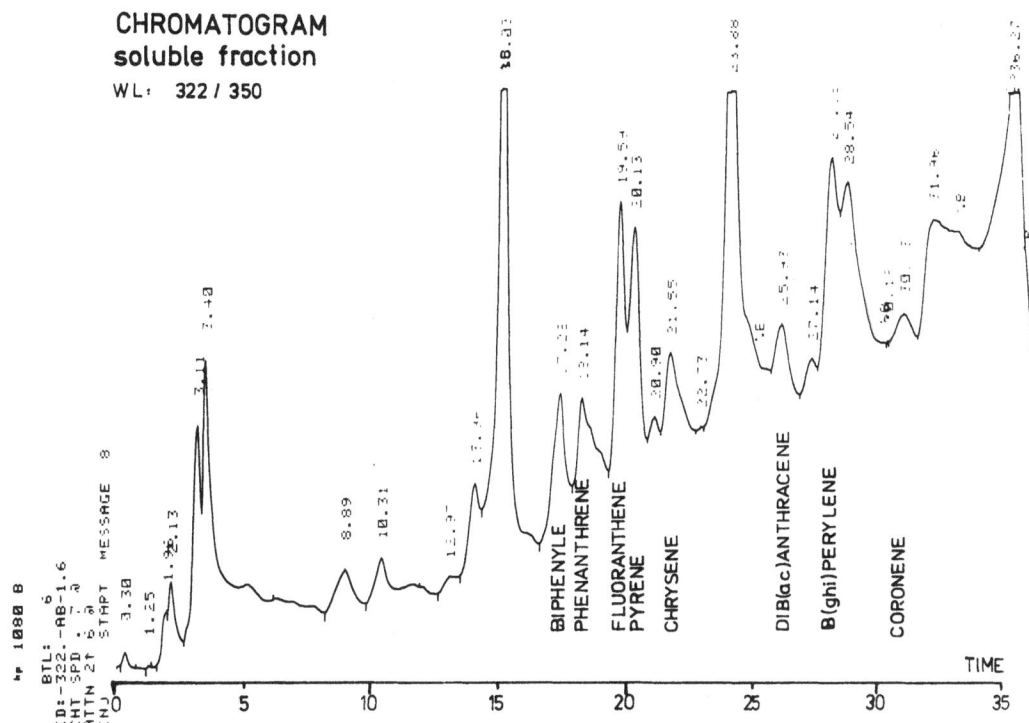

Fig. 1: Chromatogram of PAH in precipitation
(soluble fraction)

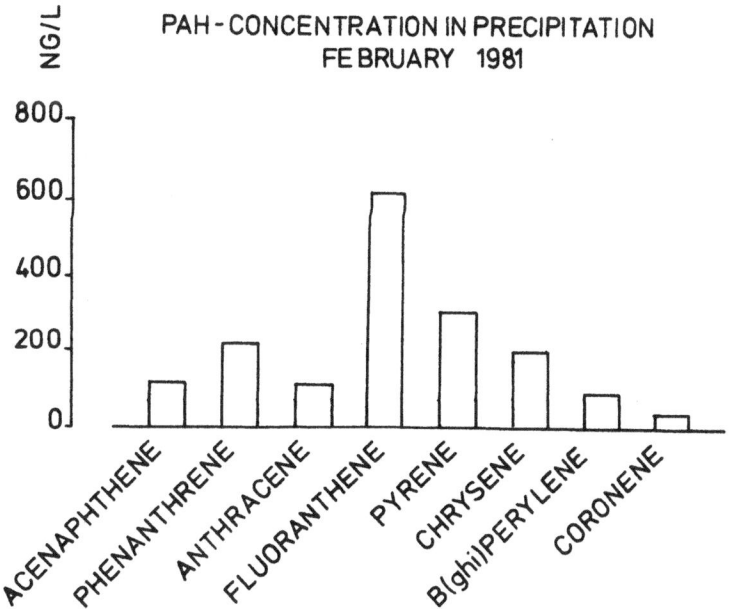

Fig. 2: Distribution of PAH-concentration
(monthly mean value February 1981)

PAH-DISTRIBUTION IN PRECIPITATION

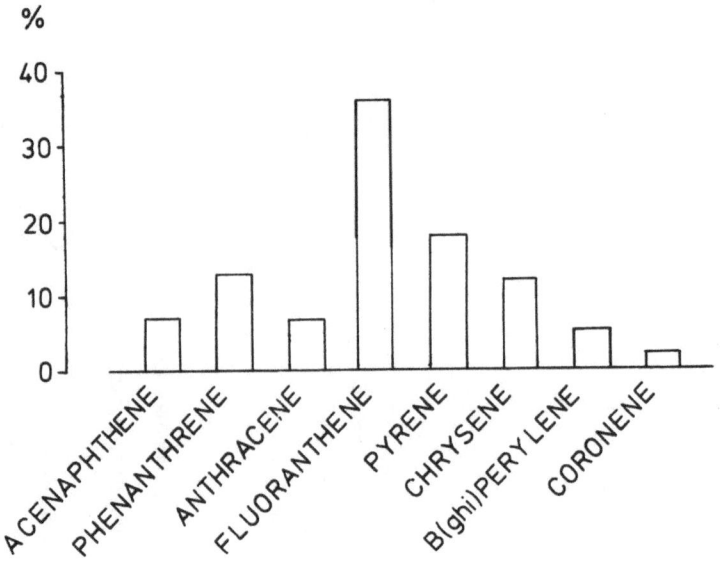

Fig. 3: Percentage distribution of PAH (monthly mean value February 1981)

PAH DISTRIBUTION SOLUBLE / INSOLUBLE FRACTION

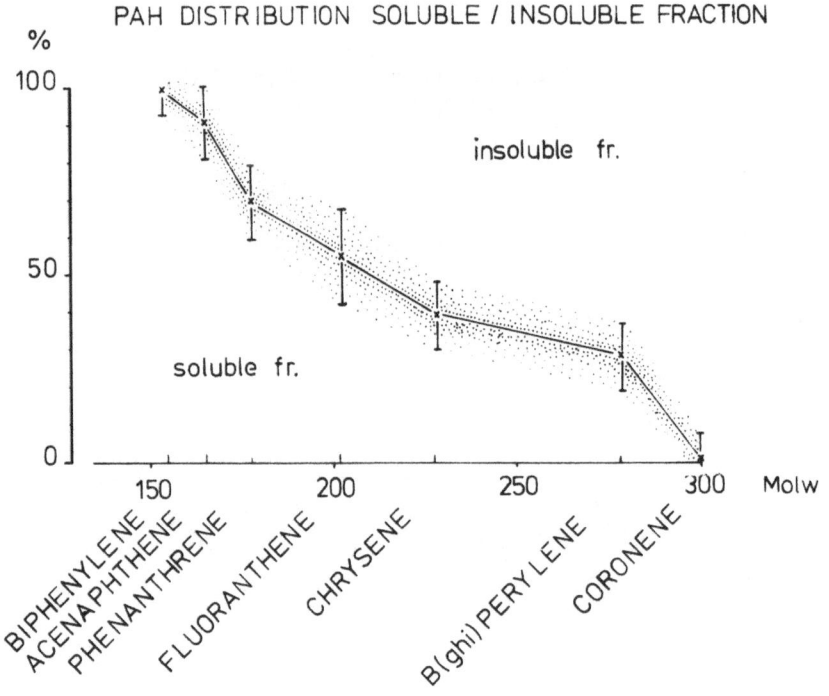

Fig. 4: Percentage distribution of the soluble and insoluble fraction

PAH-components in precipitation is shown once again. Flou-
ranthene is the main constituent of the total PAH load,
while the other components Pyrene, Phenanthrene and Chry-
sene are of less importance. The remaining compounds Ace-
naphthene, Anthracene, Benzo (ghi)perylene and Coronene
are of minor importance in the total PAH-load in precipitation.
Nearly the same percentage distribution of the PAH-compounds is found in all precipitation samples. The absolute
concentrations or the meteorological conditions do not influence the characteristical PAH-profile. The very similar profiles result in an uniform PAH-immissionprofile in
the atmosphere (4). In order to control the evaluation
method for routine determinations, the constant profile may
be used to define one single compound, which represents the
indicator for the whole PAH-spectrum in precipitation.

Ratio: soluble fraction - insoluble fraction. Fig. 4
shows the composition of the soluble and insoluble fraction
in precipitation. The individual components show (according to their molecular weights) a different distribution
in the soluble and the insoluble fraction. The soluble fra-
tion decreases with increasing molecular weight of the components.
The distribution of the soluble and insoluble fractions depends on the extension of the dry period before the precipitation. After a long dry period, the insoluble fraction increases strongly and most of the total PAH load (65%) is
found in the insoluble fraction, while in sequential samples
of individual rainfalls the insoluble fraction decreases
(35%). In snow samples, the insoluble fraction is higher
(70%) due to increased adsorption on the snowflakes.
The different ratios between insoluble and soluble fraction
in precipitation are characterized by the different behavior of the PAH-compounds. One effect is the solubility
of PAH in H_2O (Fig. 5). The solubility of PAH with low molecular weights lies in the range of μg/l. With increasing
molecular weight, the solubility of PAH in water decreases
to ng/l (5). Both curves in figure 5, the solubility and
the measured ratio between the insoluble and soluble fraction show a parallel course.
Under atmospheric conditions, PAH-compounds with lower molecular weights were found in particulate matter as well as
in the gas phase (6). With increasing molecular weight,
PAH-components are adsorbed on particulate matter and yield
only a small contribution to the soluble fraction. Decrease in water solubility and decrease in gas phase, therefore, lead to the measured ratios.

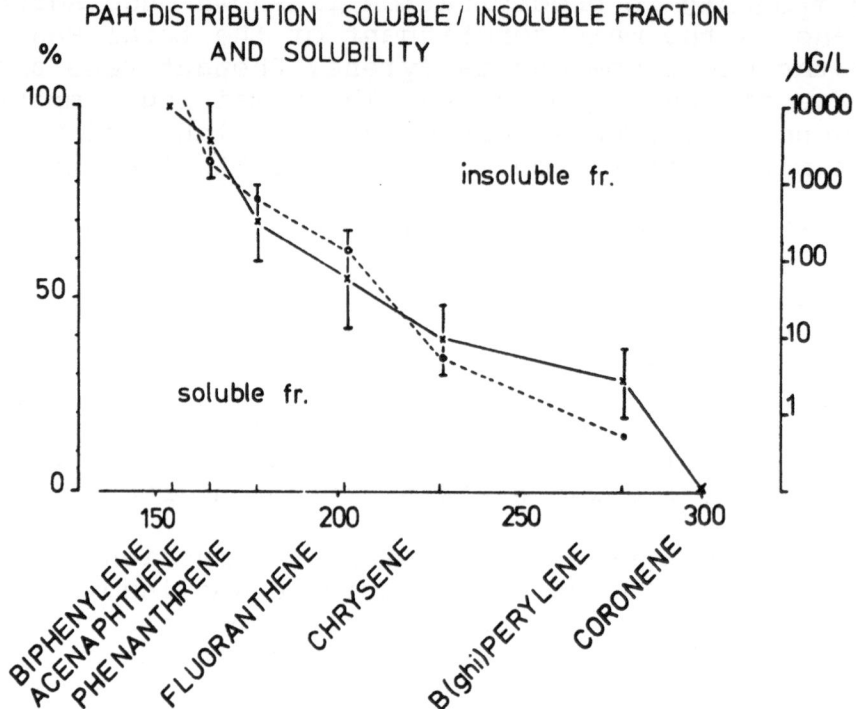

Fig. 5:   Percentage distribution of the soluble
          and insoluble fraction and solubility
          of PAH in H_2O

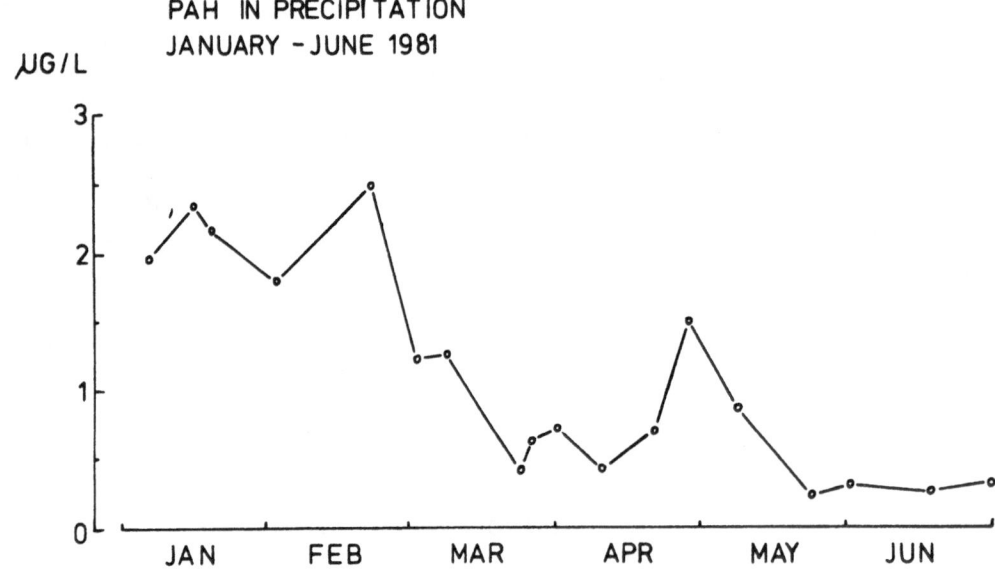

Fig. 6:   Seasonal variation of the total PAH load
          in precipitation from January-June 1981

SEASONAL VARIATION

The seasonal variation of PAH in precipitation is shown as the sum of all measured PAH-compounds from January - June 1981 (Fig. 6). In winter, there is a significant maximum with highest concentration of about 2,4 µg/l. The average values range from 2,0 - 2,2 µg/l. With the beginning of the warmer months the concentrations decrease rapidly. Then the total PAH load is in the range of 1 µg/l, while in summer only a tenth of the winter values were measured, i.e. 0,2 - 0,3 µg/l. The seasonal variation demonstrates a periodicity with high concentrations in winter, decreasing concentrations in spring, and low concentrations in summer. This behaviour is determined by additional effects. The temperature shows an inverse course to the seasonal concentrations (Fig. 7). High PAH-concentrations are found at low temperatures, while increasing temperatures result in a decrease of the concentration. At high temperatures, the processes of oxidation by atmospheric trace gases (NO_x, SO_2, O_3) is more efficient, i.e. in summer, the degradation rates are higher than in winter. The same periodic variation with temperature show some specific reactants (O_3) are higher in summer than in winter.
The distinct seasonal trend is effected by the change of the emissions with reduced domestic heating during summer. Estimations of the contribution of domestic heating result in more than 35% of the whole global PAH-emission (7).

REGIONAL DISTRIBUTION

The numerous snowfalls during winter 1980/81 allowed to determine the regional distribution of PAH in the Rhein-Main-area. From individual snowfalls, samples from the upper layer of newly fallen snow were taken using purified glass plates, then packed air-tight in bags and stored at temperatures of < -10°C, until they were analysed in the same procedure as the rain samples.
In the Rhein-Main-area the regional distribution of the PAH show high concentrations in the urban area and decreasing concentrations at the suburban stations (Fig. 8). The highest concentration of nearly 6 µg/l was measured in the city, while the concentrations at the suburban stations (3 µg/l) are reduced by a factor 2. The lowest concentration was found at a distance of 25 km from the city on Mt. 'Kleiner Feldberg' in 800 m altitude as well as in the north of the town in a large park. Therefore, a significant gradient between the city stations and the suburban region is observed. Furthermore, there exists a different composition of the soluble and insoluble fractions of the samples taken at the various stations (Fig. 9). While the total load of PAH in

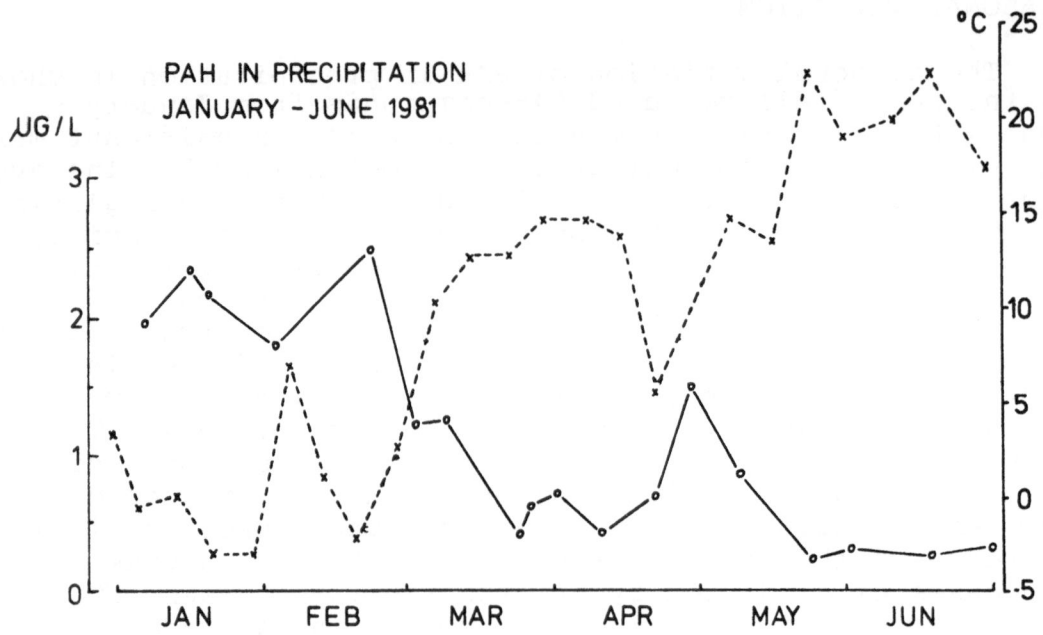

Fig. 7: Seasonal course of temperature and PAH-
        concentration from January–June 1981

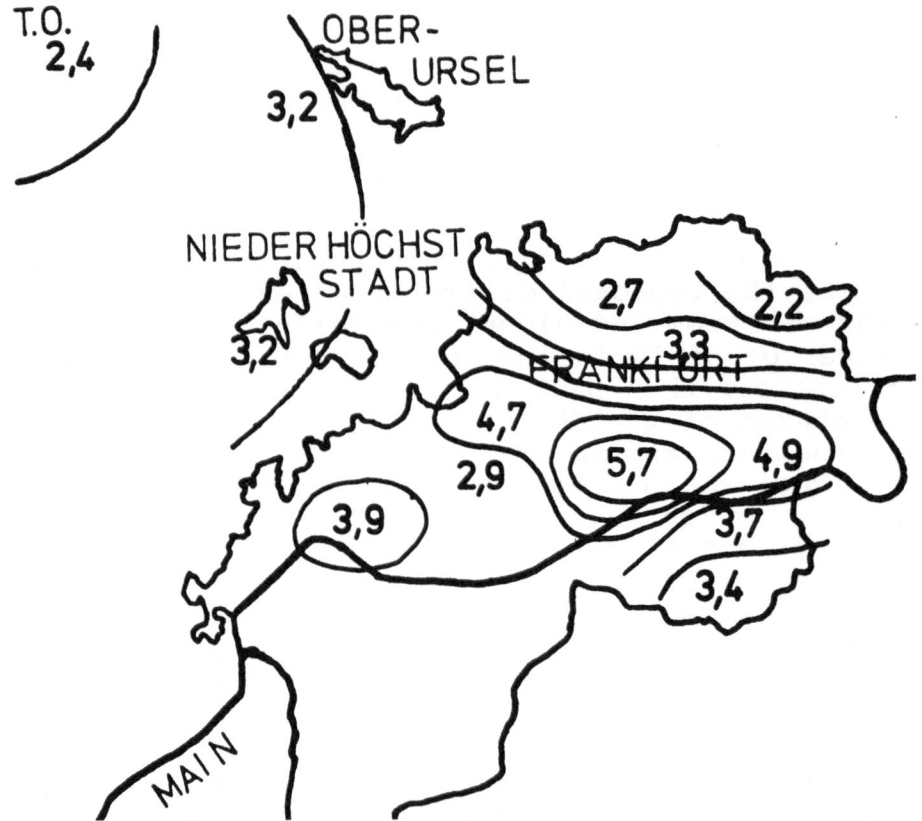

Fig. 8: Regional distribution of total PAH in pre-
        cipitation in the Rhein-Main-area, units ug/l

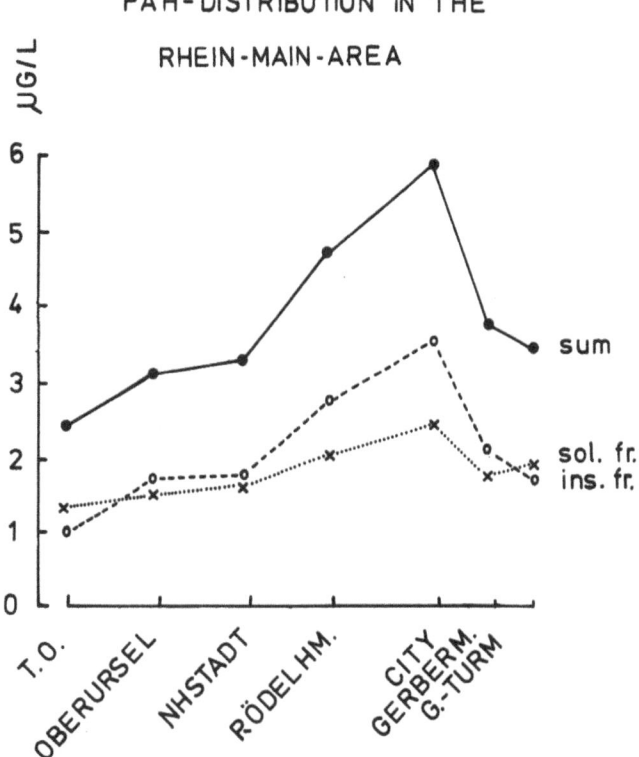

Fig. 9:   Concentrations of soluble and insoluble
          fraction at the various stations

precipitation in the city shows an intensive increase, the
individual contribution of the soluble and insoluble fractions do not increase in the same way. In comparison to
the suburban stations, where insoluble and soluble fractions
have similar concentrations, the insoluble fractions in the
city are much higher than the soluble fractions. Therefore,
the regional distribution of PAH in precipitation shows
directly the influence of the local sources in the urban
region.

ACKNOWLEDGMENTS

The author wishes to thank Dr. J. Müller for his helpful
discussions and ideas on the subject and the Umweltbundesamt, Pilotstation Frankfurt/Main for the permission to use
the instruments.

REFERENCES

(1)   Pierce, R.C. and M. Katz (1975)   Dependency of PAH
      content on size of atmospheric aerosols.
      Environ. Sci. Technol. 9 347/353

(2)   Pott, F. and R. Dolgner ((1979)   Polyzyklische aro-
      matische Kohlenwasserstoffe.
      Staub Reinhalt. Luft 39, 443/452

(3)   Müller, J. and Rohbock, E. ((1980)   Method for measure-
      ment of PAH in particulate matter in ambient air.
      Talanta 27, 673/675

(4)   König, J. et al. (1981)   Untersuchung von 135 PAH in
      atmosphärischen Schwebstoffen aus 5 Städten der
      BRD.
      Staub Reinhalt. Luft 41   73/78

(5)   May, E. (1978)   Determination of the solubility be-
      haviour of some PAH in water.
      Anal. Chem. 50 997/1000

(6)   Neff, J. M. (1978)   PAH in the aquatic environment.
      Applied Science Publisher LTD London

(7)   Suess, M. J. (1976)   The environmental load and cycle
      of PAH.
      Total Environ. 6   239/250

WET DEPOSITION OF TOXIC METALS FROM THE ATMOSPHERE IN THE
FEDERAL REPUBLIC OF GERMANY

H.W.Nürnberg, P.Valenta, V.D.Nguyen
Institute of Applied Physical Chemistry, Chemistry Department,
Nuclear Research Center (KFA), Juelich, Federal Republic of
Germany

ABSTRACT

Toxic metals are a class of environmental chemicals of great significance and high priority in ecotoxicology. In Germany practically only anthropogenic sources are responsible for their emission to the atmosphere from which the investigated metals are brought predominantly by wet deposition onto the vegetation blanket. A network of 16 automated samplers, collecting specifically wet precipates, has been distributed over the Federal Republic. The wet deposition of Pb, Cd, Cu, Zn and Se(IV) during 198o has been determined subjecting the samples to simultaneous metal analysis by differential pulse stripping voltammetry. The deposition data are discussed in context with the pollution sources in the regions and locations of the sampling sites. Findings of general significance reveal, that always over 9o % of those toxic metals are dissolved in the rain and are consequently deposited as species most favourable for uptake by the vegetation. During the initial 2 h of rainfalls the toxic metal concentration is, due to intensive wash out, substantially higher than the more or less steady-state levels in the remaining period of rainfalls. Additional pH-measurements provide also an overview on the acidity pattern of the rain in 198o.

INTRODUCTION

Certain heavy metals and metalloids constitute a class of environmental chemicals with great significance and priority in ecochemistry and ecotoxicology (1-4). This refers to the for man, mammalians and many other organisms a priori toxic metals Cd, Pb and Hg and also to a number of metals and metalloids (e.g. Cu, Zn, Ni, Se, Cr, As) which turn above respective thresholds for the respective organism type from essential functions or not hazardous levels into toxic actions. Metals and metalloids are characterized by special ecochemical features. They are not biodegradable but undergo a biogeochemical cycle during which only transformations into more or less toxic species occur (3,4). They have also the tendency to be accumualted by organisms. Therefore, they exert

143

H.-W. Georgii and J. Pankrath (eds.), Deposition of Atmospheric Pollutants, 143–157.
Copyright © 1982 by D. Reidel Publishing Company.

progressively increasing chronical toxic actions in man and mammalians
according to the levels of long term exposure. For man and mammalians
the most important common source of uptake is food. For Pb also respira-
tive uptake can be significant in areas with heavy automobile traffic.
One of the most significant pathways of toxic metals from anthropogenic
and natural emission sources to the entrance points of the food chains
goes through the atmosphere from which the toxic metals are brought by
dry and predominantly by wet deposition into the terrestrial and aquatic
ecosystems. In this context it has to be emphasized that the input of
toxic metals dissolved in rain water constitutes a particular hazard,
because it provides the metals in a form which is most favourable for
their uptake by the vegetation blanket and inland waters. Moreover, the
in our latitudes usually acid rain will leach and dissolve an additional
amount of metals from dust particles which had been brought before by
dry deposition onto the vegetation. The acquisition of a sufficiently
comprehensive number of reliable data on the toxic metal input from the
atmosphere by wet deposition is therefore an urgent prerequisite to
clarify the situation of the environmental burden created via the amtos-
pheric pathway in the Federal Reepublic of Germany. As contribution
to this task the institute has set up a network of automated samplers
(see Figure 1) which monitors since 1980 the wet deposition in various

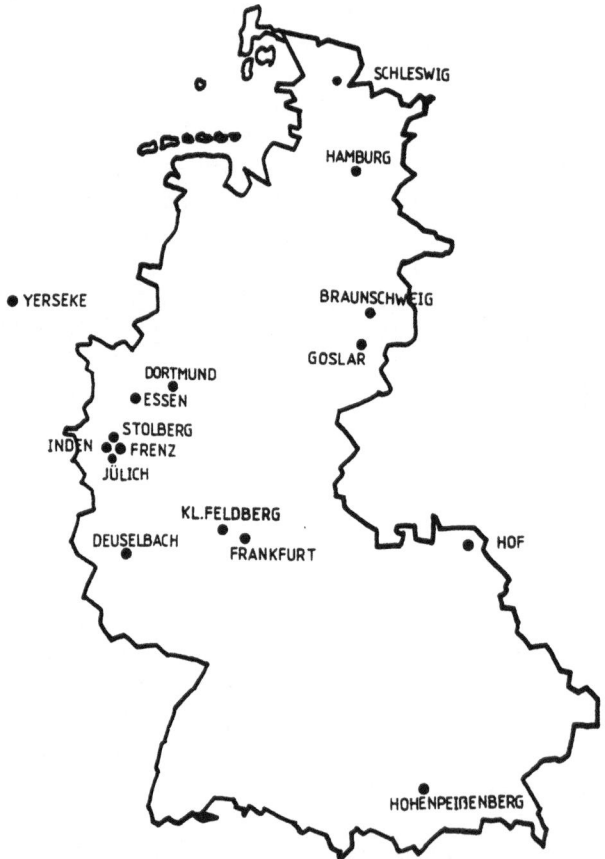

Figure 1. Sampler network

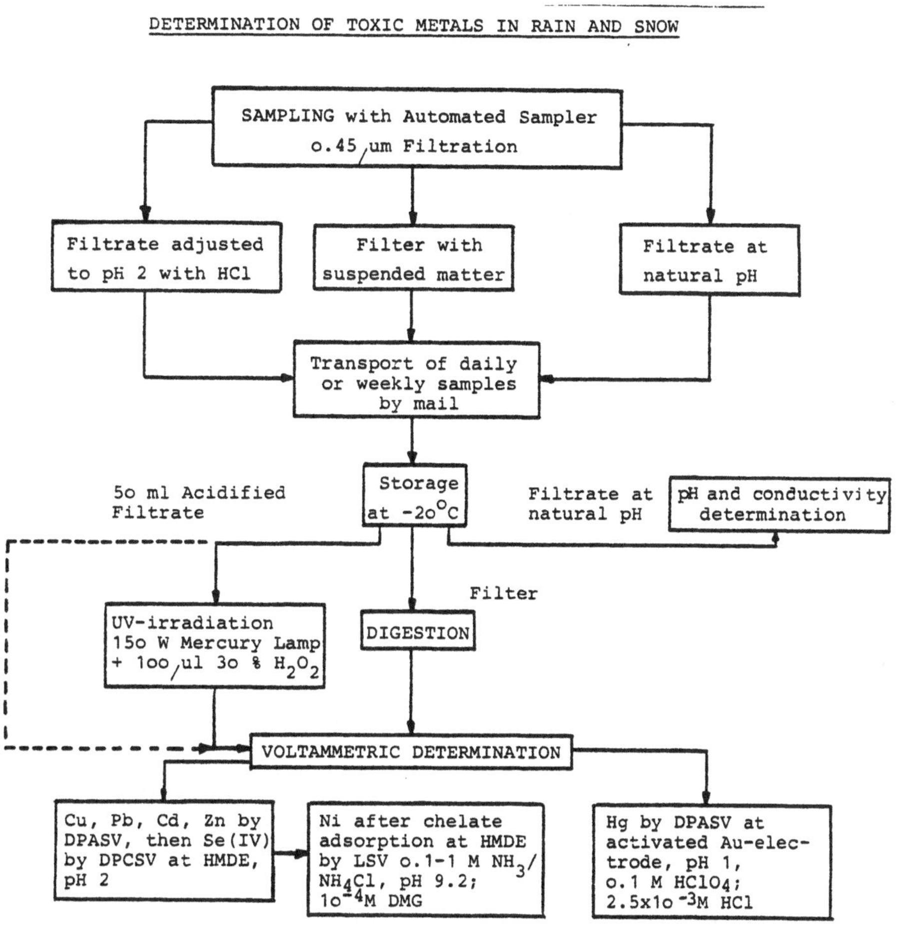

DETERMINATION OF TOXIC METALS IN RAIN AND SNOW

Figure 2. Flow chart of analytical procedure

types of ecosystems (coastal zones, rural regions, urban agglomerations and areas of heavy and metallurgical industry). In Germany toxic metal emission to the atmosphere occurs practically from anthropogenic sources only (5). The most significant emissions originate from coal burning, iron and steel production, metallurgical industry and garbage incineration. A particular case is Pb where automobile traffic contributes about 6o % of the total emission.

EXPERIMENTAL

The decisive prerequisite for the meaningful investigation of toxic metal input by wet deposition are as in other branches of ecochemistry reliable and accurate analytical data (6,7). In this respect rather demanding problems of trace metal analysis have to be overcome satisfactory avoiding particularly analytical contamination errors in all stages of the analytical procedure from sampling until determination. The chosen analytical procedure is depicted in fig. 2.

For wet deposition sampling an automated sampler (8) was constructed

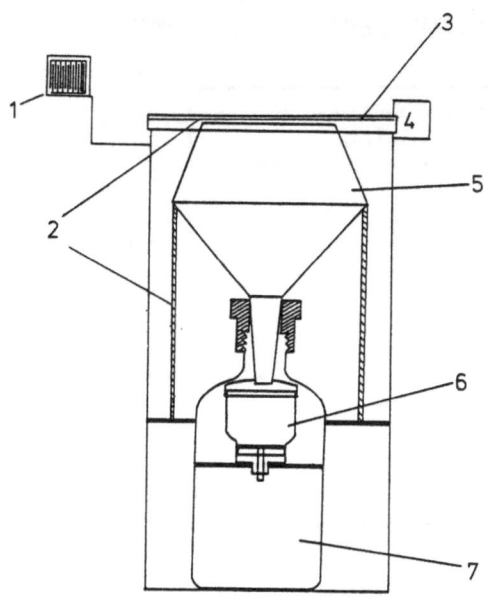

Figure 3. Automated sampler. 1 - humitidy sensor; 2 - heating
          elements; 3 - cover; 4 - servo mechanism; 5 - funnel;
          6 - filtration device; 7 - sampling bottle

(see Figure 3). Controlled by a humidity sensor it opens and closes at
the begin and end of rain- and snowfalls within an adjustable interval
of o.5 to 3 min. Proper adjustment of this time span prohibits malfunc-
tion due to mist, fog and seaspray. In this manner only wet deposition is
collected and interference by dry dust deposition is excluded (9,1o). The
position of the sampler on a rack 2 m above ground prohibits interferen-
ces by splashing of soil. By the ambient temperature triggered heating
ensures operation down to $-3o^{o}C$ and melts eventually sampled snow. The
core of the sampler is a polyethylene flask having on top a filtration
device, Sartorius SM 16511, with a o.45 µm membrane filter and a collection funnel also made from polyethylene. In this manner suspended matter
is filtered off from the collected precipitates. These components of
the sampler are subjected before use to scrupulous cleaning procedures,
now common in aquatic trace metal chemistry (11), to keep accuracy
risks due to blank values negligibly small. The production costs of
this sampler are with about 15oo $ moderate.

According to local service facilities either daily or weekly samples
are collected and transported to the institute where they are eventually
stored until analysis. In an aliquot of the filtrate pH and via conductivity ionic strength are determined before storage.

In further aliquots of the filtrate the toxic metals are determined by
appropriate modes of voltammetry (6,7,1o,12,13). In this context it has
to be emphasized that in terms of reliability, sensitivity, costs and
convenience the electroanalytical approach of voltammetry has proved to
be the method of first choice for the analysis of toxic metals in all

kinds of natural waters and provides also in combination with appropriate digestion a very significant analytical alternative for toxic metals in other matrices (6,7).

For the determination of toxic metals dissolved in rain and snow usually acidification of the filtrate suffices as pretreatment. Only in certain heavily industrialized areas (e.g. Ruhr region) the content of dissolved organics in the rain water required additional pretreatment (11) of the sample by UV-irradiation under addition of H_2O_2 to decompose by oxidative photolysis organic material which tends to bind strongly a part of the dissolved heavy metals.

The solution is deaerated with 99.999 % N_2 for 1o min and Cu, Pb, Cd and Zn are determined simultaneously by differential pulse anodic stripping voltammetry (DPASV) at the hanging mercury drop electrode (HMDE) (14). At first the metals are preconcentrated as amalgams in the mercury drop by cathodic deposition at -1.2 V (SCE) under stirring (9oo rpm) for 2 to 5 min. Then the stirring is terminated and after a quiescent period of 3o s the metals are determined by anodic stripping in the differential pulse mode using the following settings: pulse height 5o mV, pulse duration 35 ms, clock time of pulses o.3 s, scan rate of mean electrode potential 5 mV/s. Subsequently Se(IV) is determined by differential cathodic stripping voltammetry (DPCSV). Therefore Se is preconcentrated on the electrode as a HgSe-film by adjusting for 3 min a deposition potential of -o.2 V and scanning subsequently the potential in the differential pulse mode to more negative values. In this manner the Hg(II) in the before formed HgSe-film is reduced. The evaluation of the concentrations of the beforementioned metals is performed by two standard additions. The determination of the 5 metals takes thus about 3o min. For a more detailed description of the voltammetric measurements see refs. 1o,12-14. While in 198o only the aforementioned metals have been determined meanwhile in a less comprehensive number of samples Ni and Hg are determined in addition (15). This is achieved for Ni in the same analyte after pH-change by addition of ammonia buffer and interfacial acummulation by adsorption of the Ni-chelate formed with the added chelator dimethylglyoxime (DMG) (16). For Hg a separate aliquot has to be measured by DPASV at the gold electrode (17) (see Figure 2).

In the same manner the voltammetric determination of the metal content in the filtered off suspended matter can be performed as well in the analyte resulting from the prior digestion (9,14).

All manipulations of the samples during the pretreatment and determination stages are carried out under laminar flow clean benches (class 1oo) excluding in this manner contamination risks from the laboratory atmosphere (11). The detection limits at a precision corresponding to a RSD $\leq$ 1o % are given in Table 1. At higher concentrations the precision is even better. This suffices fully for the analysis of atmospheric precipitates and the detection limit could be easily lowered by a factor of 2 - 3 if a RSD $\leq$ 2o % is tolerated.

Table 1. Detection limits (ug/l) with RSD ≤ 1o % of differential pulse stripping voltammetry at HMDE

Cu	Pb	Cd	Zn	Se(IV)
o.5	o.1	o.o5	o.5	o.2

RESULTS AND DISCUSSION

The results on the wet deposition of Cu, Pb, Cd, Zn and Se(IV) obtained
during 198o at 15 sampler stations in the Federal Republic and the station Yerseke in the Scheldt Estuary, Netherlands, (see Figure 1), are
presented and discussed. In addition the pH of the precipitates has been
followed (see Figure 8).

At all stations always over 9o % of the total amount of the with the
wet precipitates deposited metals have been found in the dissolved
state in the samples. This confirms in a more general manner based on
a substantially larger data number for the investigated metals earlier
observations obtained during previous years already for certain regions
(9,14). There occurs obviously efficient wash out by dissolution of the
metals in the rain. The rather significant acidity indicated by a frequent average pH ≤ 4.5 (see Figure 8) will be favourable in this respect.

This effect of rain on the fate of the studied toxic metals in the atmosphere provides an explanation on the predominant significance of
their wet deposition compared to their dry deposition. This predominance of wet deposition is further emphasized for these metals from the below reported time pattern of their concentration in the rain during precipitation periods. In this context it has to be emphasized that there
are according to the findings of other authors (18) other heavy metals
for which the distribution between wet and dry deposition is much more
balanced or even alternative. But this is not the case for the with
respect to ecotoxicity particularly important metals investigated in
our study, at least not in our latitudes. From these findings follow
two further conclusions of general significance.

For environmental monitoring purposes it will usually suffice to determine in our latitudes for the studied toxic metals only the dissolved
metal amounts to control with respect to environmental protection their
wet deposition.

Of particular ecotoxicological significance is the fact that wet deposition being the prevailing deposition mode for those toxic metals
brings them almost exclusively onto vegetation in a form most favourable for uptake and consequently for introduction into the terrestrial
and aquatic food chains. Moreover, the acidity of the rain will favour
in addition leaching and subsequent uptake of those metals from dust
particles brought by dry deposition onto the vegetation.

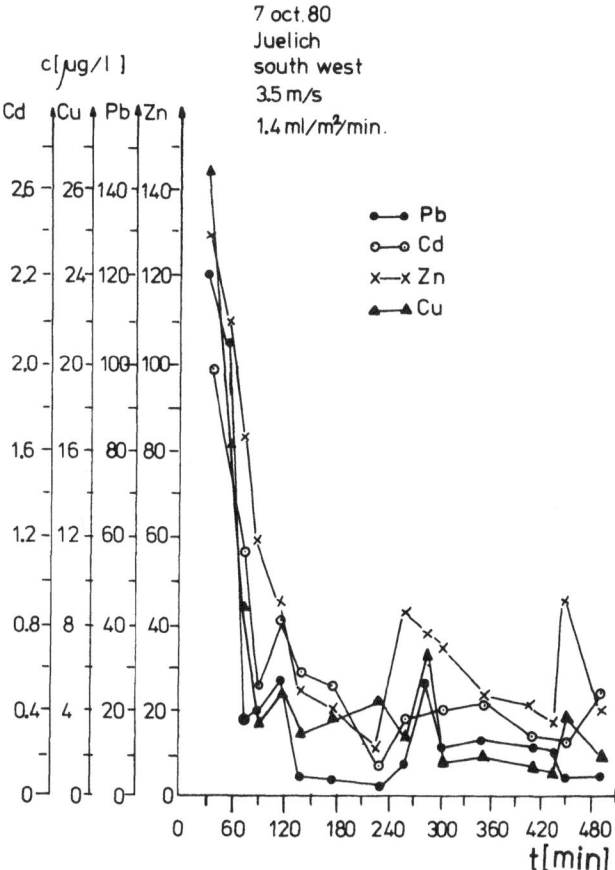

Figure 4. Typical example for time function of metal concentration in rain water during rainfall. Juelich
(stat.5), 7.1o.198o; precipitation: 1.4 ml/m²min;
wind SW, 3.5 m/s.

In extended studies, which are treated in more detail elsewhere (19),
also the time pattern of the concentration of the studied toxic metals
n rain has been followed over certain rain periods. A typical example
is shown in fig. 4. The samples have been taken with a second type of
our automated sampler equipped to sample over short time intervals
samples in a fractionated way. The fact that the chosen voltammetric
ᵈₑtermination method requires only rather small volumes (5 ml) for a
measurement is very convenient for those investigations. The time
pattern of the metal concentration has always a similar trend. The
concentration is significantly higher in the initial phase, i.e. by
a factor 7 or more, than in the stationary phase attained after about
2 h of a rainfall. This indicates that the metal pollution assembled
in the atmosphere due to emission before the rainfall is washed out to
a large extent in the initial phase. Then a more or less steady state of
wet deposition is reached as the regional emission and the contribution from the long distance transport within the atmosphere to the
sampling site continue during the remaining period of the rainfall. The

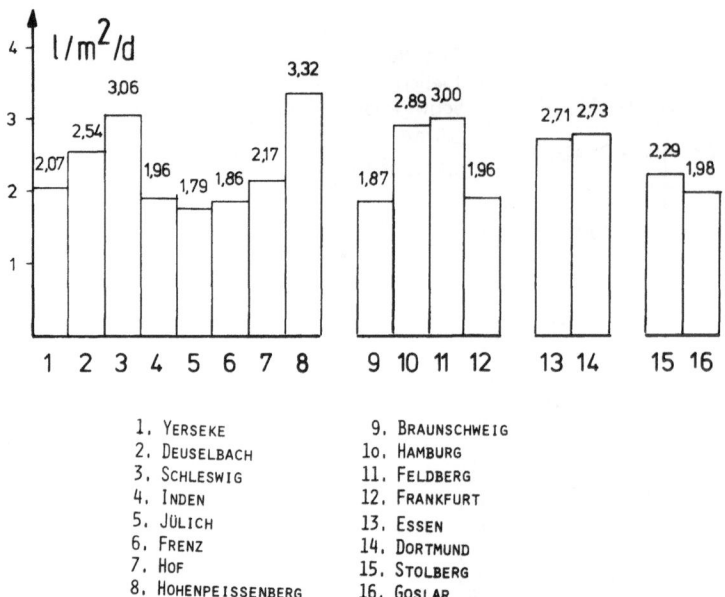

Figure 5. Average daily wet precipitation in 198o at sampler
stations

steady state levels of the metal concentrations in the rain may undergo,
however, some fluctuations, due to corresponding fluctuations in the
regional emission and/or meteorological conditions, particularly wind
direction and wind strength, which cause corresponding alterations in
the amount of atmospheric transport of toxic metals from longer distances towards the sampling site. In general the observed time pattern
underlines again the predominance of wet deposition among the deposition modes for the studied toxic metals.

In fig. 6 and 7 the average daily wet deposition of the toxic metals
and in fig. 5 the average daily wet precipation for the sampling stations of our network in 198o are given.

Average Daily Wet Precipitates

In fact the measured actual values during the rain periods of the year
are higher. From these here with respect to space not shown data the
average values have been computed by averaging the resulting annual
input over 365 d. Obviously for the average wet deposition of metals
the average daily wet precipitation is of primary significance.

Although this average daily wet precipitation was in 198o of comparable
magnitude for all stations, the difference between the lowest value at
Braunschweig (stat. 9) and the highest value at Hohenpeissenberg (stat.
8) amounts to 1.53 $1/m^2$, i.e. by about a factor of 2. One can distinguish
several groups, i.e. stations 1, 4, 5, 6, 7, 9, 12, 15 and 16 with a daily
average of wet precipitates between 1.8 and 2.3 $1/m^2$; stations 3, 8, 1o,
11, 13, 14 with 2.7 to 3.3 $1/m^2$ and station 2 with 2.54 $1/m^2$. In the

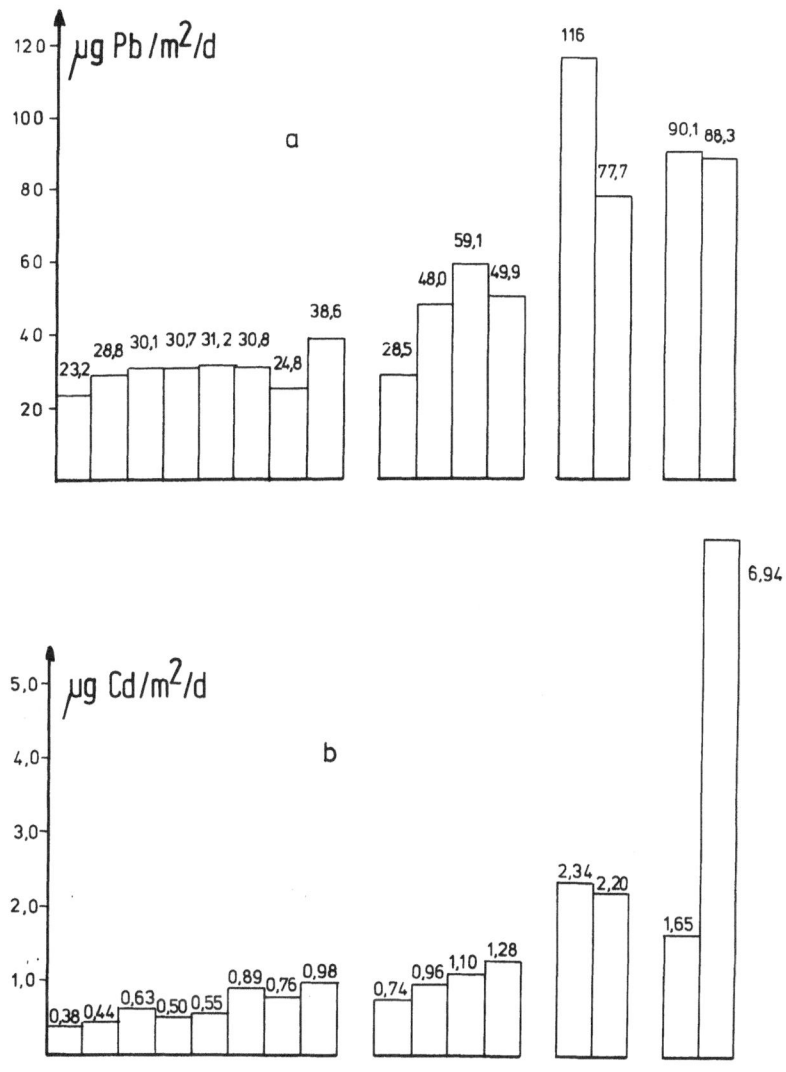

Figure 6. Average daily wet deposition in 1980 at sampler
stations. a) Pb; b) Cd

group with the lower daily average precipitation fall, including Deuselbach (stat. 2) with 2.54 $1/m^2$, all stations in coastal and rural areas
where particular definite emission sources in the immediate surrounding,
i.e. within a circle of 15 to 20 km, are absent but to this group belong
also stations 12, 15 and 16, where either a pronounced regional area of
emission sources (Frankfurt, station 12) is active or well defined and
strong point emission sources, due to metallurgical industry (Stolberg,
station 15, and Goslar, station 16). In the second group with significantly higher average daily wet precipitation fall stations affected by
the influence of efficient regional emission source areas in urban agglomerations (stat. 10, 11) or heavy industry areas (stat. 13,14) as well
as two stations (3, 8) from rural regions without particular emission
sources.

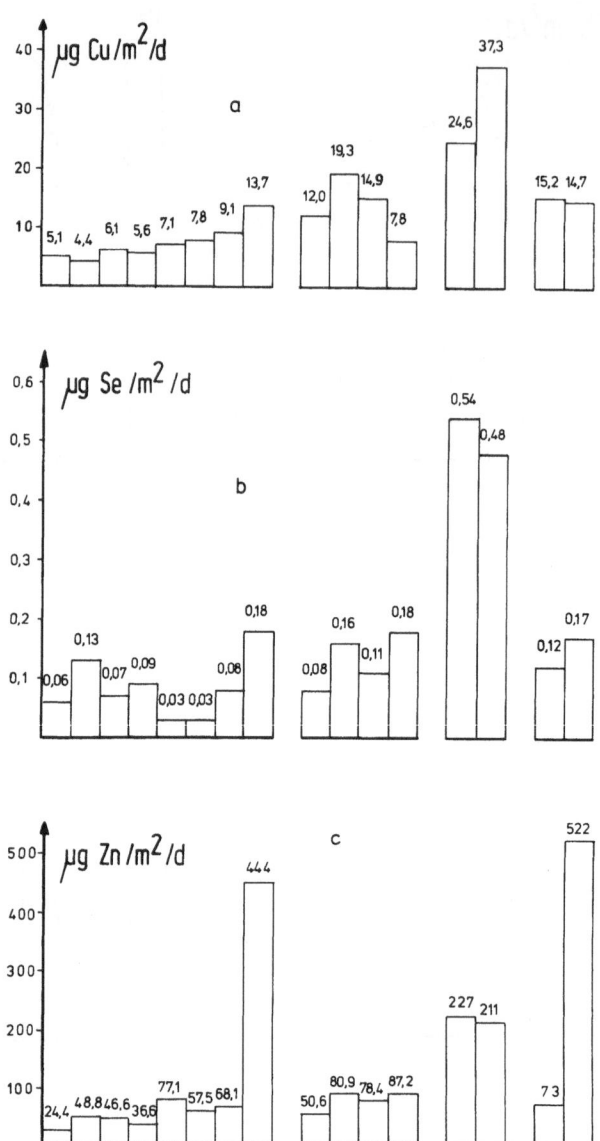

Figure 7. Average daily wet deposition in 198o at sampler
stations. a) Cu; b) Se(IV); c) Zn

Wet Deposition of Toxic Metals

Of paramount importance for the wet deposition of the respective toxic
metal is, of course, the presence of emission sources in the surrounding of the sampling sites and the efficiency, number and type of those
sources. This is well reflected by the results.

For Pb (see Figure 6 a) within a factor 2 the deposition is similar for
stations 1 to 12 with a distinct tendency for the higher values at
stations 1o to 12, affected by urban agglomerations, compared with the
lower Pb-deposition at the stations in rural or coastal regions (stat.
1 to 9).

The data for Stolberg and Goslar (stat. 15 and 16) reflect the influence of Pb-smelters acting as definite sources, but the highest value has Essen (stat. 13) being under the accumulated influence of heavy automobile traffic, typical for urban agglomerations, and the area emission of a number of industrial sources (iron and steel industry) and coal burning installations. In this context it has to be noticed that usually a substantial amount of toxic metals emitted from a source is again deposited in its vicinity, typically within a circle of about 2o km (19). There exists a gradient of decreasing deposition with increasing distance from the source. Only a smaller amount of the emitted toxix metals undergoes atmospheric transport over long distances. The different fate of emitted toxic metals depends in their relative quantities strongly on the type of the emission sources, the type of the emitted species and the respective meteorological conditions during and after emission. A certain, often small but in its total by no means negligible amount of toxic metals is transported very far in our hemisphere and deposited in very remote otherwise unpolluted regions as the Arctic Ocean (2o). None of our stations is located close to a definite point emitter or emission area. Therefore, the determined data reflect for stations located in regions with definite point sources (stat. 15, 16) or numerous area sources (stat. 1o - 14) the average pollution with toxic metals caused by their wet deposition from the atmosphere in the respective region. For no station the average daily wet Pb-deposition has exceeded in 198o the threshold value of 25o ug/m^2d regulated as tolerable for orchards, vegetable cultivation or fish farming (21). Definitely the situation in the closer vicinity of efficient emission sources and under unfavourable meteorological conditions (inversion situations) will be substantially worse than the average deposition data tell.

For Cd (see Figure 6b) a more or less systematic increase of the wet deposition is observed increasing from coastal over rural sites without special emitters followed by urban and industrial agglomeration regions and reaching a concerning peak value at Goslar due to the action of special emitters (Zn-smelters).Although the emission sources are 5 km distant from the station, already the daily average wet Cd-deposition exceeds the according to the regulation for agricultural areas tolerated threshold value of 2.5 ug/m^2d by almost a factor 3, while for Essen and Dortmund (stat. 13,14) the average wet Cd-deposition remains just below the tolerable threshold value (21).

For Cu and Se(IV) (see Figures 7a,b) only both sampling sites in the Ruhr region, Essen and particularly Dortmund, show compared to the other stations significantly elevated although ecotoxicologically uncritical average deposition values. Here station 14 reflects by the deposition of Cu and Se(IV) the influence of one of the largest copper plants in Europe located within a distance of 5o km but in the over the year prevailing wind direction towards the sampling station at Dortmund.

For Zn (see Figure 7c) usually values between about 5o and 9o ug/m^2d are observed with a trend to the upper end of the range in urban agglomerations and stations affected by them (stat. 1o, 11, 12), but also

Juelich (stat. 5) shows a higher value than common rural areas. Particularly low is the value at Yerseke (stat. 1) where the prevailing wind
comes from the sea. The stations in the Ruhr region (stat. 13, 14)
have more than a factor 2 higher values and Goslar (stat. 16) has as
expected the peak value. Interesting is also the unusually high deposition at Hohenpeissenberg (stat. 8). It is caused by a local point emission source as which a zinking plant located in the prevailing wind
direction in 5 km distance acts. As it is not a zinc smelter the Cd-deposition, however, remains within the common levels (see Figure 6b).

pH of Rain

As an additional parameter having also impact on the toxic metals in
precipitates the pH of rain has been measured. Fig. 8 represents the
distribution of pH measured in rain during 1980 at the stations of the
network (19). It should be noticed that for the majority of the stations the pH has been measured in a sample collected over one week.
Only in Deuselbach, Juelich and Dortmund (stat. 2, 5, 14) pH was determined in daily samples. Nevertheless, the data clearly reflect a distinct trend for the frequency of pH towards more acid values in correlation with the increasing regional air pollution. The primary reason
for the increased acidity of the rain above the natural background value
around 5.6, adjusted by the carbonate buffer system in clean areas (e.g.
Pacific Ocean (14), Arctic Sea (20)), is the more or less strong air
pollution with SO_2 and NO_x in our latitudes. The relatively lowest acidity is observed for the with respect to local emission sources relatively unpolluted coastal and rural regions (stat. 1 - 6); even here pH
is usually below 4.7 reaching sometimes one unit lower and rather frequently at least up to 0.5 units more acid values. Of course, the pH-
frequency pattern for each station has its own individual characteristic
distribution leading either to a widely spread out pH-range (Juelich
stat. 5; Hohenpeissenberg, stat. 8) or a pH-frequency more focussed to
the range 4.2 - 4.5. At Hof (stat. 7) obviously special effects of certain air pollution sources cause a relative acid range (3.9 - 4.4) for
the pH-frequency. These local special pollution influences make the pH-
frequency at this with respect to toxic metals rather low affected station more comparable with the pH-frequencies of stations (10, 12, 13,
15, 16) with increased air pollution, including toxic metals, due to
either manifold emission sources active in urban agglomerations or specific industrial sources. Remarkable is also the pH-pattern at stat.8
(Hohenpeissenberg) with two pH-frequency maxima, around 4,0 and between
4.4 and 4.7 suggesting that under certain meteorological conditions
additional air pollution contributions become effective at this sampling
site. The most significant example for by rather acid pH reflected air
pollution is Dortmund (stat. 14), where incidentally the pH-frequency
pattern is based on daily pH-determinations during the rain periods.

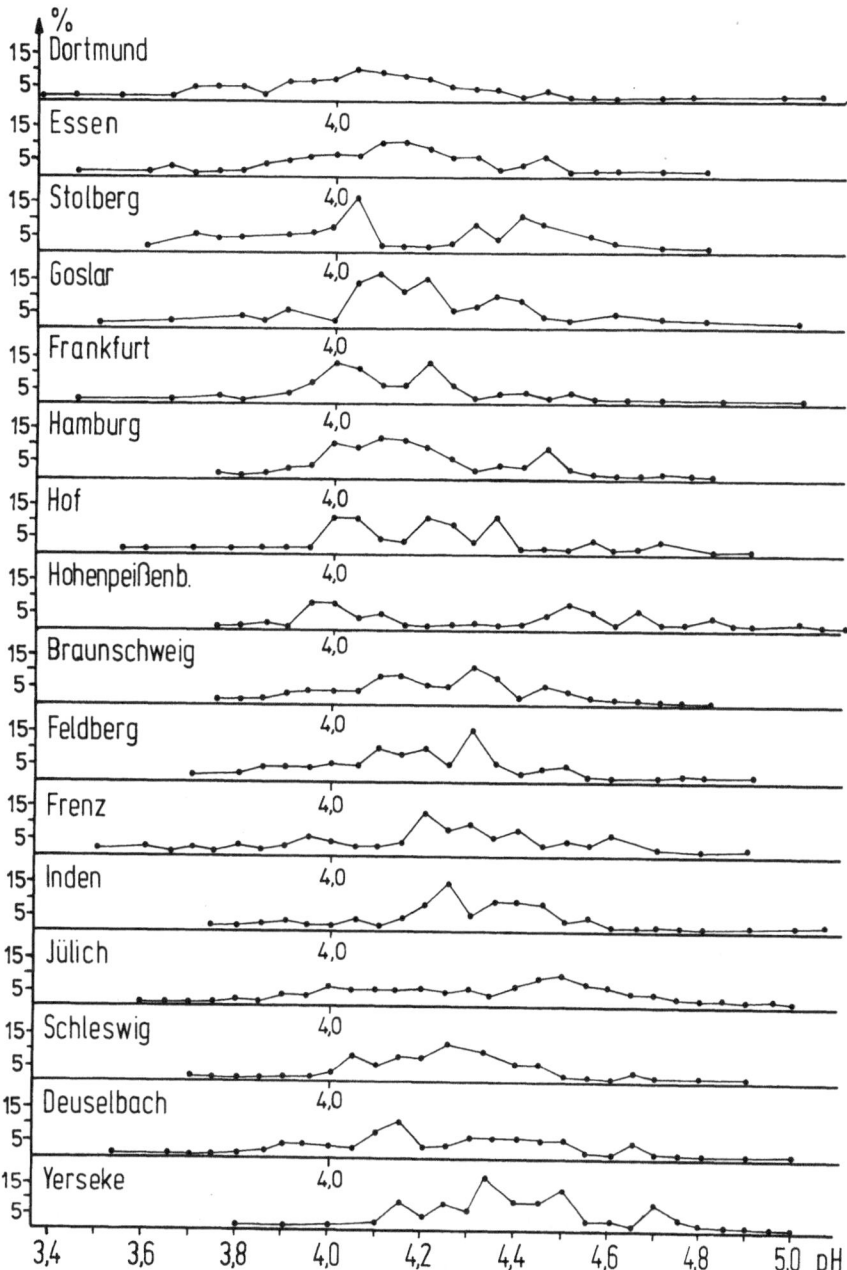

Figure 8. pH-frequency of rain in 1980 at sampler stations

CONCLUDING REMARKS

The in 1980 with the over the Federal Republic of Germany distributed sampler network obtained and evaluated data on the wet deposition of the studied toxic metals and on pH of rain had in the first place explorative character. In this respect they have perhaps posed more and better formulable questions than provided detailed final answers. Never-

theless, the investigation has shown, that a rather elucidating overview on general trends and patterns of toxic metal pollution by wet
deposition and on acid rain distribution in the various regions and ecosystem types of Germany is obtainable in this manner. Certain points of
general ecochemical and ecotoxicological significance in context with
the in its manifold details but not in its general behaviour very complicated and complex problem could be clarified relying on a substantial
basis of several thousand reliable experimental data. The going on continuation of this research program will not only yield data on the
seasonal dependencies and long term trends of the respective ecochemical
problems but is also expected to contribute significantly to deepen
the understanding on the magnitude and significance of the ecological
impacts of toxic metal and acid rain deposition.

ACKNOWLEDGEMENTS

The authors thanks V. Eicker and H. Salgert for skillful technical
assistance.

The sampler network is operated with the gratefully acknowledged tech-
nical support of German Meteorological Service (stat. 3, 7, 8, 9, 13);
Institute of Meteorology and Geophysics, Univ. Frankfurt (stat. 11, 12);
Analyt. Chemistry, Univ. Dortmund (stat. 14); German Fed. Env. Agency
Station Deuselbach (stat. 2); City of Goslar (stat. 16); Delta Institute,
Yerseke (stat. 1). Further thanks go to German Meteorological Service
for continuous supply of meteorological data and with respect to stat.
4, 5, 6 and 15 by the meteorological group of KFA.

REFERENCES

1. Venugopal, B.; Luckey, T.D., 1978, Metal Toxicity in Mammals, Vol.1,2,
   Plenum Press, London-New York.
2. Friberg, L.; Nordberg, G.F.; Vouk, B., 1979, Handbook on the Toxico-
   logy of Metals, Elsevier/North Holland Biomedical Press, Amsterdam
3. Merian, E.; Geldmacher-v.Mallinckrodt, M.; Machata, G.; Nürnberg, H.W.;
   Schlipköter, H.; Stumm, W.; Hrsg., Metalle in der Umwelt, Verlag Chemie,
   in press
4. Hutzinger, O., 1980, Environmental Chemistry, Vol. 1-3, Part A,
   Springer, Berlin-Heidelberg-New York
5. Schladot, J.D.; Nürnberg, H.W., 1982, Atmosphärische Belastung durch
   toxische Metalle in der Bundesrepublik Deutschland - Emission und
   Deposition", Institut f. Angew.Phys.Chemie, Kernforschungsanlage
   Jülich, 73 pp.
6. Nürnberg, H.W., 1979, Sci. Tot. Env. 12, pp. 35-60
7. Nürnberg, H.W., Pure Appl. Chem., in press
8. Nguyen, V.D.; Valenta, P., 1978, German Potent Applic. 2831 8403
9. Nürnberg, H.W.; Valenta, P.; Nguyen, V.D., 1979, KFA-Jahresbericht,
   pp. 47-54
10. Nürnberg, H.W., 1979, Chem.Ing.Techn. 51, pp. 717-728
11. Mart, L., Talanta, in press

12. Nürnberg, H.W., 1981, Analytiker-Taschenbuch, Vol. 2, pp.211-23o, Springer, Berlin-Heidelberg-New York

13. Valenta, P.; Nürnberg, H.W., 198o, Gewässerschutz-Wasser-Abwasser, 44, pp.1o5-2o1

14. Nguyen, V.D., Valenta, P.; Nürnberg, H.W., 1979, Sci.Tot.Env. 12, pp. 151-167

15. Gödde, M., Urano de Carvalho, E.F.; Valenta, P.; Nürnberg, H.W., Sci.Tot.Env., to be submitted

16. Pilhar, B.; Valenta, P.; Nürnberg, H.W., 1981, Fresenius Z.Anal. Chem., 3o7, pp.337-346

17. Sipos, L.; Golimowski, J.; Valenta, P.; Nürnberg, H.W., 1979, Fresenius Z.Anal.Chem. 298, pp.1-8

18. Gravenhorst, G.; Perseke, C.; Rohbock, E., 198o, "Untersuchung über die trockene und feuchte Deposition von Luftverunreinigungen in der Bundesrepublik Deutschland", Inst.Meteorologie u. Geophysik, Univ. Frankfurt, 52 pp.

19. Nguyen, V.D.; Valenta, P.; Nürnberg, H.W., Sci.Tot.Env., to be submitted

2o. Mart, L.; Nürnberg, H.W.; Dyrssen, D., "Low Level Determination of Trace Metals in Arctic Sea Water and Snow by Differential Pulse Anodic Stripping Voltammetry" in Wong, C.S., ed., "Trace Metals in Sea Water", Plenum Press, New York, in press

21. Verwaltungsvorschrift zur Änderung der Ersten Allgemeinen Verwaltungsvorschrift zum Bundes-Immissionsschutzgesetz, Deutscher Bundestag, 8. Wahlperiode, Drucksache 8/2751

ATMOSPHERIC REMOVAL OF AIRBORNE METALS BY WET AND DRY DEPOSITION

Eberhard Rohbock
Institute for Meteorologie and Geophysics,
Frankfurt/Main

ABSTRACT

At 13 sites of different air quality in the Federal Republic of Germany dry and wet deposition was determined separately. Over the period summer 1979 to summer 1981 chemical analyses were performed for lead, cadmium, manganese and iron by atomic absorption spectrometry.
Mass balances of wet and dry deposition rates show the different significance of both removal pathways. Elements bound on large aerosol particles (Mn and Fe) are removed from the atmosphere preferentially by dry deposition, whereas elements bound on submicron particles are removed by wet deposition.
To qualify both removal processes deposition velocities and washout factors are calculated. Characteristic deposition velocities of the metals confirm the sedimentation as the dominant mechanism. Characteristic washout factors are in the range from 100 to 5000. A correlation between deposition velocities and washout factors demonstrates that dry deposition and wet removal are most efficient for metals bound on large particles. Nevertheless, wet deposition is the main removal process for submicron particles.

INTRODUCTION

Heavy metals in the atmosphere are existent predominantly in form of aerosol particles. The gaseous phase is of minor importance for total mass. The aerosolbound metals are chemically inreactive. Therefore, atmospheric removal of metals is restricted to wet removal by incorporation in the precipitation and dry removal by direct deposition onto the ground. Junge (1963) distinguishes two pathways how aerosols are incorporated into precipitation water. The first way called "rainout" describes the incorporation within the cloud itself. Condensation of

159

H.-W. Georgii and J. Pankrath (eds.), Deposition of Atmospheric Pollutants, 159–171.
Copyright © 1982 by D. Reidel Publishing Company.

watervapour on aerosolparticles particularly on hygroscopic
and watersoluble particles is the dominant process. Below
the cloud base the falling raindrops and snowflakes capture aerosol particles due to different sedimentation velocity. This mechanism is called "washout". The analysis of
precipitation water at the ground makes it nearly impossible
to differentiate the ways of scavenging.

EXPERIMENTAL

    In the scope of the German deposition network (Georgii
et al. 1980a) dry and wet deposition rates of various atmospheric trace substances have been measured by an automatic wet/dry deposition collector. The collector has been
installed at 13 sites of different air quality over the
Federal Republic of Germany. The sites are characterized
in detail by Perseke (1981). Deposition measurements have
been carried out over a 2 year period from summer 1979 to
summer 1981.

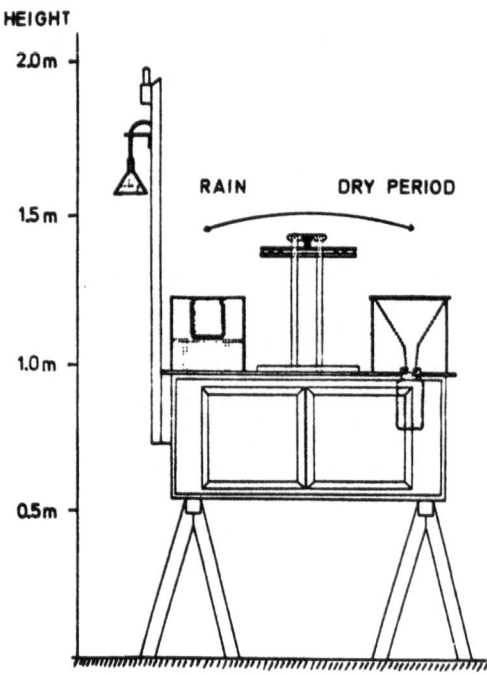

Fig 1.: Wet/dry deposition collector

The wet/dry deposition collector used in the deposition
network is shown in Figure 1. The two sampling funnels -
one for precipitation water, the other for dry deposition -
are covered by a lid alternatively. A detailed description
of the collector is given by Georgii et al. (1980a). Precipitation water is collected in polyethylene bottles

(1 liter) through a 25 cm ⌀ polyethylene funnel. Dry deposition matter is collected in highwalled glass containers according to the standard method of Bergerhoff (VDI-Richt-linie 2119, 1972). The glass containers have been proved to be most suitable for collecting dry deposition in field measurements as well as for the chemical analyses of various elements (Rohbock et al. 1981). Simultaneously with the deposition measurements airborne metals are enriched on two filters fixed at a height of 1.5 m above ground.

Sampling periods of dry deposition are 14 days. The rain samples are taken daily. Measuring periods of aerosol concentration are 7 days. The samples collected at the different sites are posted to the central laboratory at the Meteorological Institute, Frankfurt/Main where measurements of pH-value, conductivity and heavy metals are performed. In order to distinguish between the water-soluble and water-insoluble fraction of the metals, the samples are filtered through membrane-filters (0.45 μm poresize). Dry deposited matter and the coated aerosolfilters are washed with de-ionized water by aid of an ultrasonic bath beforehand. Chemical analyses by atomic absorption spectrometry include the toxic metals lead, cadmium and iron, manganese. For certain periods, dry deposition and aerosolfilters are analyzed on Cu, Ni, V, Co, Cr, Na, K, Mg, Al, Si, additionally.

RESULTS AND DISCUSSION

For the valuation of the relative importance of dry and wet deposition the relative frequency of dry and wet periods must be taken into account. The temporal distribution of wet periods during the 14 day sampling periods are demonstrated in Figure 2. Generally, wet periods - when the raingauge was opened - amount to less than 20% of the 14 day periods as shown by the example of Braunschweig in Figure 2. At Essen, the amount of wet periods increases to about 30 to 40%. These extended rainy periods at Essen have to be explained by the orographic situation of the sampling site situated south of Essen. Though the wet periods are of longer duration at Essen, the total rainfall amounts for the periods (Fig. 5) are not elevated. This means that these periods are characterized by low rain intensities.

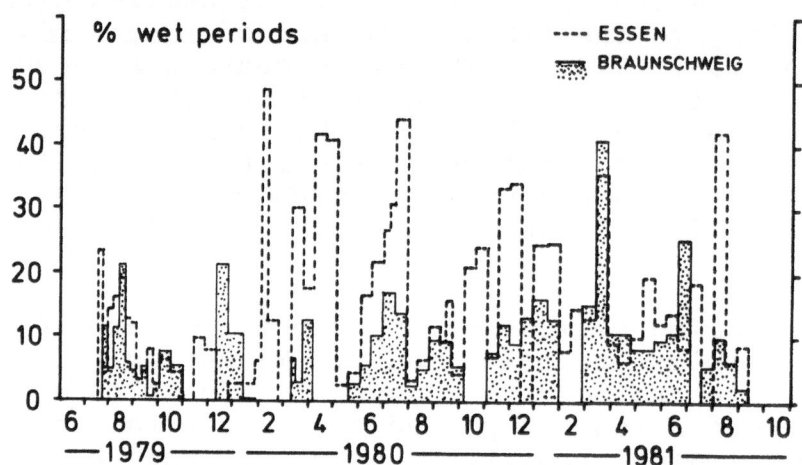

Fig. 2: Fraction of wet periods during the ex-
position periods of dry deposition vessels

METALS IN PRECIPITATION WATER

The investigation of the water-soluble and water-insoluble fractions of the metals allows a first assessment of
the bioavailability of the metals. Water-soluble metal compounds especially lead and cadmium can be absorbed by live-
stock. On the other hand, the knowledge of the distribution of metals in precipitation water is essential for as-
sessing the data of some authors who analyzed only the water
soluble species of some metals (Nürnberg, etal., 1981).

In Figure 3, the comparison of the water-soluble fractions
of lead, cadmium, manganese and iron in rainwater are given
in form of a cumulative frequency distribution. Lead, cadmium and manganese are found in soluble forms predominately.
The average values of insoluble fractions of cadmium and
lead (50% values) account to less than 10%; the relevant
values for manganese are about 12%. Nevertheless, in a-
bout 10% of the rainsamples the insoluble fraction of manganese and lead exceeds 30%. In contrast to these soluble
metals, the iron distribution in precipitation water shows
higher fractions of insoluble compounds. The average value
amounts to 43%. In 40% of the samples the insoluble fraction predominates. The different distribution of iron is a
first indication of a different behaviour of iron which will
be discussed later on.

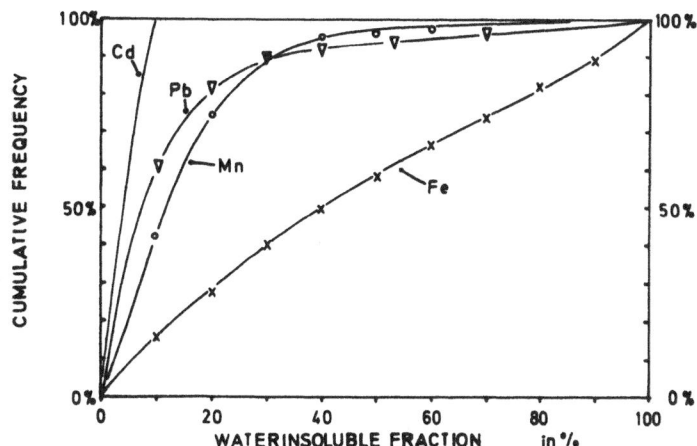

Fig. 3:   Distribution of the water-insoluble
          metals in precipitation water

The distribution in Figure 3 gives mean conditions in considering samples of all sampling sites.   There exist small
differences of the soluble/insoluble distribution of metals
in rainwater sampled at stations of different air quality.
In polluted and urban areas, the insoluble fractions of lead
and iron are higher than in rural and clean areas.   The mean
fraction of insoluble lead compounds in rain sampled at
Frankfurt-city is 22.5%, whereas on the Taunus-mountains,
25 km NW of Frankfurt (823 m altitude), the insoluble fraction accounts to 13.9%.   A similar but less expressed trend
is found for iron.   An explanation of this trend may be
washout of aerosols in urban areas.   Washout predominately
removes coarse particles of the lower layers.   (These particles are less soluble than aerosol particles of the submicron range . Müller,1981).

The absolute metal concentration in rainwater are determined by various parameters, like rain intensity and direction of advection.   Nevertheless, local influences are
obvious by comparison of the concentration found at Essen
and Deuselbach, as in Figure 4.   Essen represents a polluted site at the southern border of the Ruhr-district,
whereas Deuselbach is representative for unpolluted air.
At both sites the metal concentrations vary over 2 orders
of magnitude at least.   Concentrations are elevated at
Essen compared to Deuselbach by a factor 2 to 4 for all
metals.   The deltoides give the range of concentration.
The peaks represent the maximum and minimum values whereas
the average values (50% values) are indicated by the cross-
beam.

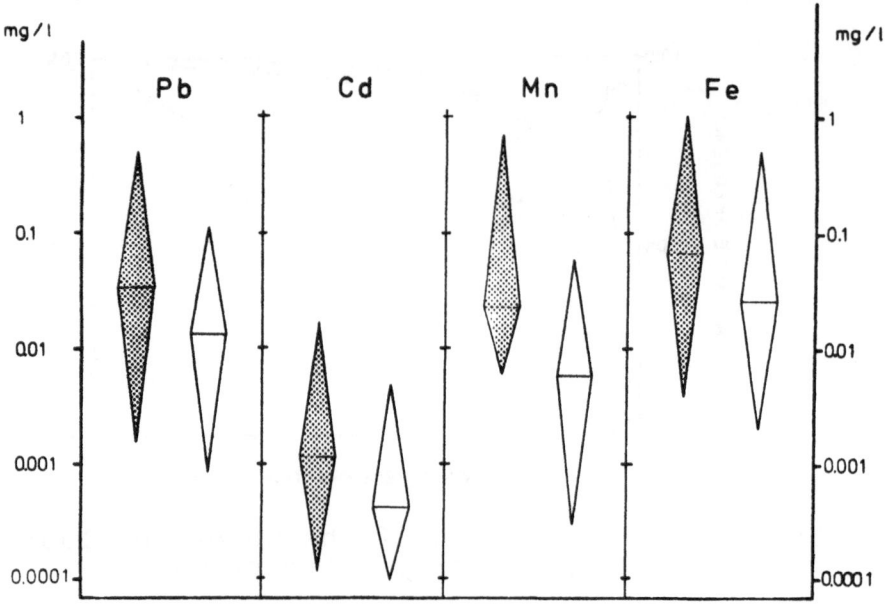

Fig. 4:   Metal concentration in precipitation wa-
          ter collected in a clean area (Deuselbach,
          white deltoides) and in a polluted area
          (Essen, pointed deltoids).

DEPOSITION OF HEAVY METALS

    Wet deposition can be calculated from the measured me-
tal concentrations in rainwater and the total rainfall a-
mount.   The results of the two year investigation suggest
that total wet deposition is determined by the precipitation pattern to a larger extend.   Highest deposition a-
mounts are always found during periods of high precipitation rates (example Deuselbach-lead in Figure 5).

LEAD

    At Deuselbach, total lead depositions are in the range
of 30 μg m^{-2} d^{-1} over extended periods.   This value has to
be considered as the background deposition of lead in the
Federal Republic of Germany.

In comparison to wet lead deposition the contribution of
dry deposition to total deposition is negligible.   In contrast to wet deposition, dry deposited amounts show a uniform temporal pattern (Figure 5).   Dry deposition is found
in a range between 5 μg Pb m^{-2} d^{-1} and 10 μg Pb m^{-2} d^{-1}.
During winter and spring seasons slightly elevated rates of
dry deposition are found.   This is in accordance with mea-

surements of total dust depositions (Lahmann und Fett,1980)

LEAD DEUSELBACH

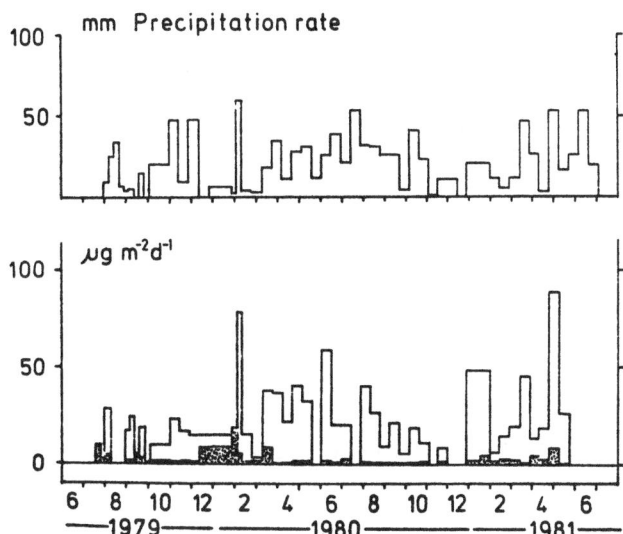

Fig. 5: Temporal trend of dry and total deposition of lead at a clean area (Deuselbach)

In industrial and urban areas the total lead depositions
increase to values up to 160 μg m^{-2} d^{-1} as a sum averaged
over a 1 year period at Essen. During single 2-week periods lead deposition rates reach values up to 300 μg m^{-2}d^{-1}.
In urban areas the fraction of dry deposited lead accounts
to about 17 to 22%, whereas in remote areas, dry deposition
contributes less than 10% to total lead deposition.

CADMIUM

Cadmium deposition shows a similar pattern. Lead and
cadmium are removed from the atmosphere by wet deposition
preferentially, although the fraction of dry deposition of
cadmium is slightly increased. In remote areas dry deposition of cadmium accounts to less than 10% of total deposition. In urban areas cadmium depositions contribute 20 to
36% of total deposition. The increased contribution of
dry deposition can be explained by gaseous cadmium. Cadmium as the most volatile species of the analyzed metals
is existent in gas-phases in the atmosphere. Total cadmium depositions were measured in the range of 1 to 4 μg
m^{-2} d^{-1}.

MANGANESE

A different deposition pattern is found for manganese and iron which are bound on coarse aerosol particles. These metals are deposited by dry deposition to a larger extend. Total manganese depositions account to 24 to 70 μg m^{-2} d^{-1}, 50 to 80% of this amount are dry deposited. At resort and mountain stations with high rates of precipitation, fractions of dry deposition are 35 to 46%.

IRON

Iron depositions are 320 to 1530 μg m^{-2}d^{-1}. Highest values are found in the Ruhr area with heavy metal-industries, as well as in Jülich where brown coal is mined by surface mining in the vicinity of the station. Dry deposition predominates to about 70% of total deposition as an annual mean.

DEPOSITION VELOCITIES - WASHOUT FACTORS

The deposition mechanisms for aerosols can be characterized by two empirical parameters - the deposition velocity and the washout factor (Chamberlain,1965). The deposition velocity is defined as the proportionality factor of measured deposited material and the air concentration. It describes the efficiency of the mechanism of transport to the ground. This parameter combines all physical and meteorological influences stability conditions, microphysical behaviour of single suspended particles and surface conditions (Sehmel,1980). According to the deposition velocity the "washout factor" is defined as the proportionality factor of the concentration of a trace substance in rainwater and the concentration in the surrounding air. High values of the washout factors show that the substances are removed by wet deposition in a efficient way. By regarding this parameter it is impossible to distinguish the pathways by which the substances initially have been incorporated into the precipitation elements.
In Figure 6, the average values of the deposition velocity of metals are summerized in dependance of the mean aerosol mass size distribution of the metals. The data show that the characteristic deposition velocities of the metals increase with increasing mass median diameter varying from 0.06 cm s^{-1} (lead) to 2,2 cm s^{-1} (magnesium). This means that the mechanism of dry deposition is 40 fold more efficient for magnesium aerosols than for lead aerosols.

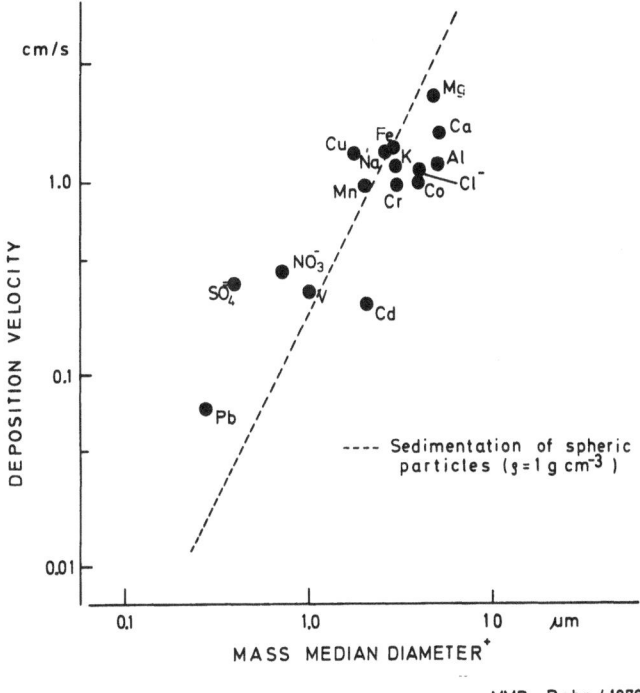

Fig. 6: Mean characteristic deposition velocities of metals measured in Germany in
comparison to mean mass size distribution

The influences of variable meteorological conditions are
not taken into account in this context. The relation between the element specific deposition velocities to the
mass mean diameter corresponds with the theoretical sedimentation of spheric particles (density 1 g cm^{-1}). This
proves that the sedimentation mechanism is the dominating
mechanism for dry deposition over long periods. During
short periods - in the present case 14 days periods - other
mechanisms especially changes in meteorological conditions
can strongly influence the deposition velocity. This is
demonstrated by the temporal trends of the deposition velocity of lead at the different sampling sites of the German
deposition network (Figure 7). During April to June 1980
the deposition velocities of lead increased significantly
at the three northern sites simultaneously. The elevated
values of up to 0.6 cm s^{-1} are about 10 times higher than
the average value. First investigations of this period
leads to the assumption that large daily variations of the
relative humidity are responsible for this increase (Rohbock et al. 1981). Weather conditions during this period
were characterized by a high pressure cell over the north
sea mainly affecting northern Germany. This results in a
distinct variation of temperature and relative humidity
during day-and nighttime. At high humidities the aerosol -

particles grow which means a more efficient deposition by
sedimentation.

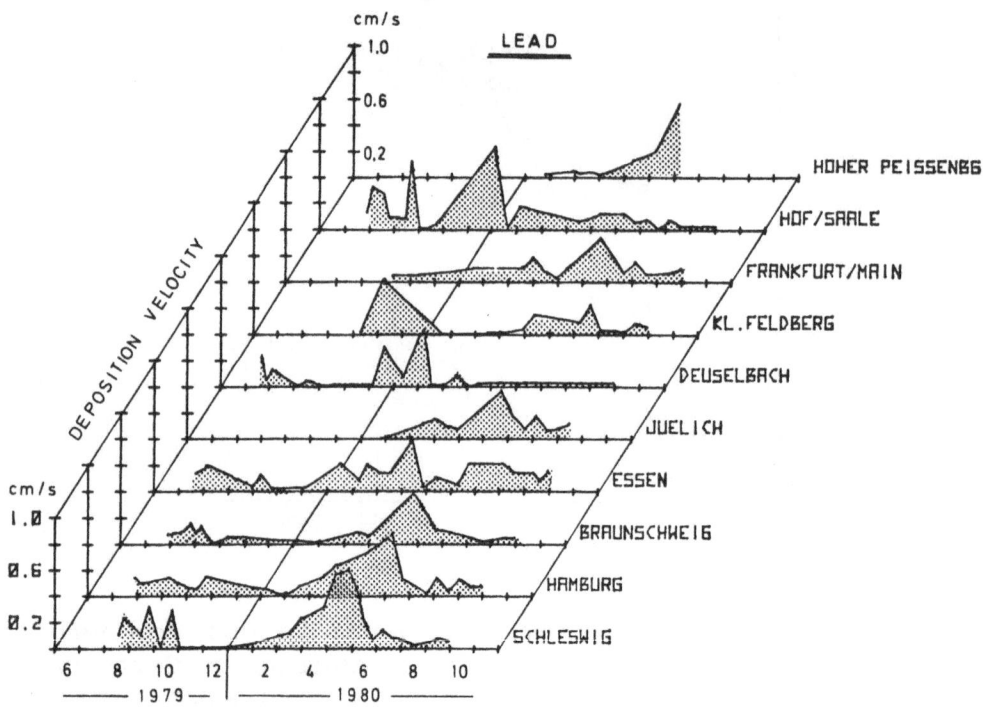

.Fig. 7:   Temporal trends of deposition velocity of
            lead measured at 10 sites in western
            Germany

In comparison with the temporal variations of dry deposition the differences of deposition velocities measured at
sites of different air quality are negligible.   This is in
agreement with the assumption that the aerosol size distribution is   conservative.

The washout factors calculated on basis of the 14 day mean
concentration in rain water and the air concentration also
revealed characteristic element-specific values.   Obviously
these values do not depend on the air quality of the stations.   At Deuselbach (resort clean area) mean values are
in the range of the sum of all stations.   The average values of the washout factors are given in Table 1.

Table 1: Characteristic washout factors of metals
The data are calculated as the quotient of concentration in rain ($\mu g\ l^{-1}$) and concentration in air ($\mu g\ m^{-3}$)

Element	Washout factors
lead	140
cadmium	400
manganese	1500
iron	70
copper	1200
calcium	1800

The table shows that manganese aerosols are enriched in precipitation water by a factor of 7 compared to lead aerosols, not considering whether rainout or washout are the dominating mechanism.

A comparison of the average factors and deposition velocity in figure 8 indicates that high washout factors are in agreement to high deposition velocities. This means that wet removal and dry deposition are more effective for metals bound on large particles. Nevertheless, wet removal especially in humid areas, still plays the dominant role in removing submicron particles from the atmosphere, while dry deposition is negligible. This is documented by the results of the mass balances of wet to dry deposited lead particles given in this paper.

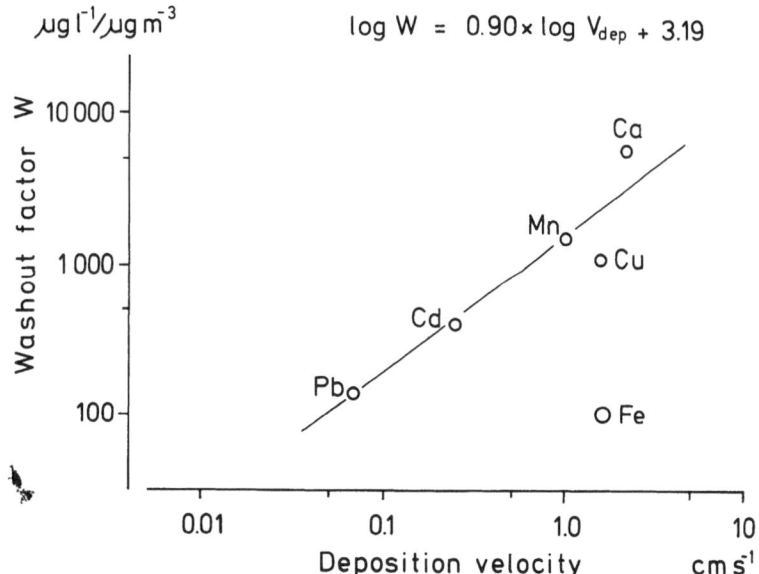

Fig. 8: Comparison of characteristic washout factors to deposition velocities of metals

Iron values do not agree to the correlation of washout factors and deposition velocities in figure 8. An explaina-
tion may be that iron aerosols are restricted to the lower
atmospheric layers which means that washout is the only removal mechanism of wet deposition. A large fraction of iron
particles is unsoluble and represent therefore less active
condensation nuclei. More research in this direction is
required.

CONCLUSION

Dry and wet deposition are pathways for the removal of
atmospheric metals. The mass balances give evidence that
metals bound on submicron aerosol particles are removed
primarily by wet deposition, whereas metals bound on coarse
particles are mainly removed by dry deposition. The accumulation of these metals is restricted to a local scale
in the vicinity of the sources. In contrast to that, lead
and cadmium which are highly toxic are distributed on a
large scale by atmospheric dispersion. Their deposition
follows the precipitation pattern. In regard of the
ecological hazard of lead and cadmium, the high solubility
of these metals has to be taken into consideration.

ACKNOWLEDGEMENT

The investigation was sponsored by the "Umweltbundesamt
Berlin" under contract number 104 02 600. The author thanks
the staff of the "Deutscher Wetterdienst, Umweltbundesamt
and Geophysikalischer Beratungsdienst der Bundeswehr" for
their technical assistance during the period of measurements. Mrs. I. Bierwerth, Mrs. M. Obeth, Mrs. H. Wallenwein are thanked for the accurate chemical analyses.

REFERENCES

Chamberlain, A. C. (1965) Aspects of the deposition of
    radioactive and other gases and particles.
    Int. J. Air Pollution 3, 63-68

Georgii, H.-W., G. Gravenhorst, C. Perseke, E. Rohbock
    (1980) Untersuchung über die trockene und feuchte De-
    position von Luftverunreinigungen in der Bundesrepublik
    Deutschland.
    Bericht - Aufträge des Umweltbundesamtes, Oktober 1980

Junge, C. E. (1963) Chemistry of precipitation, in: Air
    chemistry and radioactivity.
    Chapter 3, 289-311 , Academic Press, New York/London

Lahmann, E., W. Fett (1980) 25 Jahre Staubniederschlags-
    messung in Berlin.
    Gesundheits-Ing. gi 101 (5), 149-155

Müller, J. (1981) Größenverteilung und atmosphärische Ver-
    weilzeiten von schwebstaub gebundenen Stoffen.
    Tagungsbericht VDI Colloquium "Schwebestoffe und Stäube"
    16-18 Sept. 1981, Nürnberg

Nürnberg, H.-W., P. Valenta, V. D. Nguyen (1981) Wet de-
    position of toxic metals from the atmosphere in the
    Federal Republic of Germany, in: Deposition of at-
    mospheric pollutants, ed. H.-W. Georgii and J. Pankrath
    G. Reidel Publishing Company, 1982

Perseke, C. (1981) Composition of acid rain the the Fe-
    deral Republic of Germany - spatial and temporal vari-
    ations during the period 1979-1981, in: Deposition of
    atmospheric pollutants, ed. H.-W. Georgii + Pankrath,
    G. Reidel Publishing Company, 1982

Rohbock, E., H.-W. Georgii, C. Perseke (1981) Measure-
    ments of aerosol deposition in the Federal Republic of
    Germany, unveröffentlichtes Manuskript, Vortrag: Dry
    deposition-colloquium
    Battelle Institut, Oberursel, Mai 1981

Rahn, K. (1976) Chemical composition of the atmospheric
    aerosol.
    Technical Report, Graduate school of Oceanography,
    University of Rhode Island

Schmitt, G. (1981) Temporal and regional distribution of
    polycyclic aromatic hydrocarbons in the precipitation
    in the Rhein Main area, in: Deposition of atmospheric
    pollutants, ed. H.-W. Georgii and J. Pankrath,
    G. Reidel Publishing Company 1982

Sehmel, G. A. (1980) Particle and gas dry deposition.
    Atm. Environ. 14, 983-1011

VDI-Richtlinien 2119 (1972) Messung partikelförmigen
    Niederschlags Bestimmung des partikelförmigen Nieder-
    schlags nach dem Bergerhoff-Gerät (Standardverfahren)
    Blatt 2) VDI-Verlag, Düsseldorf, 1972

TRACE ELEMENT MEASUREMENTS IN WET AND DRY DEPOSITION AND AIRBORNE
PARTICULATE AT AN URBAN SITE

N J Pattenden, J R Branson and E M R Fisher
Environmental and Medical Sciences Division
Atomic Energy Research Establishment
Harwell, Oxfordshire OX11 ORA
United Kingdom

Samples of airborne particulate and deposited material have been
collected at a site in an industrial area of England, where a number of
ferrous and non-ferrous metal smelting and manufacturing works are
concentrated. The deposition collector automatically discriminated
between wet and dry material. The rainwater samples were filtered before
analysis in order to provide separate "rain soluble" and "rain insoluble"
fractions. The sample collection was continuous with monthly sample
changes. The samples were analysed for more than 30 trace elements by
instrumental neutron activation analysis, and for lead by X-ray
fluorescence and atomic absorption spectrophotometry.

The results are compared with similar measurements performed in rural
areas. They are discussed in terms of wet and dry deposition velocities,
particle size distributions, and the rain soluble fractions of deposition,
for various trace elements.

1. INTRODUCTION

Samples of airborne particulate and deposited material have been
collected at a site in the garden of a house in an industrial area in
an inland part of England, where a number of ferrous and non-ferrous
metal smelting and manufacturing works are concentrated. The work has
continued for 12 months, but only the first 6 months results are described here. The deposition collector automatically discriminated
between wet- and dry- deposited material. The samples were analysed for
many trace elements, including heavy metals.

The objective was to determine concentrations of trace elements in
outside aerosol and their rates and mechanisms of deposition. This
information could then be used to obtain more information about the
aerosol material, including its solubility and particle size.

H.-W. Georgii and J. Pankrath (eds.), Deposition of Atmospheric Pollutants, 173–184.
Copyright © 1982 by UKAEA

TABLE 1   Concentration in air (ng · m^{-3})
August 1980–January 1981

Element	Aug–Oct	Nov–Jan	6 month mean	Ratio to rural concentrations (a)
Antimony	50	29	40	50
Arsenic	373	291	330	130
Cadmium	72	80	76	50
Chromium	74	37	56	50
Cobalt	3	0.9	2.0	20
Copper	3410	2112	2760	200
Iron	3156	1615	2380	18
Lead	2688	3084	2880	70
Manganese	249	151	200	30
Nickel	237	45	140	20
Selenium	127	57	92	130
Silver	28	9	18	180
Vanadium	26	18	22	4
Zinc	10320	6360	8340	330
Aluminium	1068	328	700	3
Indium	1.0	0.6	0.8	10
Scandium	0.55	0.12	0.33	11
Sodium	2222	1480	1820	2

NOTE

(a) from reference 2

TABLE 2 Total deposition (wet plus dry)
Projected annual deposition rate ($\mu g.cm^{-2}.yr^{-1}$)

Element	Present measurements (a)	Range at non-urban UK sites (b)
Antimony	1.0	0.028-0.055
Arsenic	3.3	0.08-0.55
Cadmium	1.0	<1
Chromium	3.4	0.21-0.88
Cobalt	0.13	0.018-0.083
Copper	280	0.98-4.8
Iron	160	14-77
Lead	68	1.6 -4.5
Nickel	8.8	0.35-1.1
Selenium	0.65	0.022-0.065
Silver	0.25	-
Zinc	170	4.9 -12
Scandium	0.014	0.0033-0.019

NOTES

(a) Results based on separate wet and dry measurements from
August to January, with the wet measurements scaled to
projected annual rainfall of 960 mm, and the dry
measurements scaled pro rata with time.

(b) From reference (3)

2.   SAMPLING AND ANALYTICAL METHODS

The deposition collector (designed by R S Cambray, AERE) is shown in
Fig 1. In wet weather, which is detected by an electronic moisture
sensor, a plastic funnel was exposed to receive rainwater and pass it
to a plastic bottle inside the equipment. In dry weather, the funnel
was closed by a horizontal shutter on which a sheet of filter paper
(Whatman 541) was exposed and collected particulate material which
happened to fall on it. In wet weather the sheet was covered.

The rainwater samples were filtered through Whatman 42 paper before
analysis in order to provide separate "rain soluble" and "rain insoluble"
fractions.

The airborne particulate sampler consisted of a plastic holder with a
Whatman 40 filter paper, through which air was drawn at about 6 1 min^{-1}
by a pump, and the flow rate was measured with a gasmeter.

All sample collections were continuous, with monthly sample changes.

The samples were bulked into three monthly groups and analysed for about
30 trace elements by instrumental neutron activation analysis[1] to give
average concentrations in air and wet and dry depositions for each three
monthly period. In the case of lead, the samples were analysed by X-ray
fluorescence and atomic absorption spectrophotometry as individual monthly
samples without bulking.

3.   RESULTS

The measured concentrations in airborne particulate of a selected number
of trace elements between August 1980 and January 1981 are shown in
Table 1. The selection covers most of the "heavy" or "toxic" metals
plus some others. The elements As, Cu, Se, Ag and Zn are at least two
orders of magnitude higher in concentration than typical rural levels[2]
in Britain whereas the concentrations of Al, In, Sc and Na are 2-10
times typical rural levels.

The total (wet plus dry) depositions, scaled to an annual flux, are shown
in Table 2, and compared with the observed range at non-urban sites[3].
These results also demonstrate that the site is subject to considerable
local pollution by heavy metals.

The separate results on dry and wet depositions are shown in Tables 3
and 4 respectively. The rain soluble fraction of the wet deposition is
shown as a percentage.

4.   DISCUSSION

The dry deposition velocities[4], v_d, being the ratio of dry deposition
rate to concentration in air, for different elements are shown in
Table 5. The concept is extended to wet deposition velocity, v_w, with
an analogous definition, which are also shown. The values range from

TABLE 3 Dry deposition ($\mu g.cm^{-2}$)
August 1980–January 1981, and the projected annual deposition rate

Element	Aug–Oct	Nov–Jan	Projected annual rate ($\mu g.cm^{-2}.yr^{-1}$)
Antimony	0.17	0.127	0.60
Arsenic	0.62	.38	2.0
Cadmium	<0.01	<0.1	–
Chromium	0.63	.68	2.6
Cobalt	.022	.021	0.09
Copper	32.5	26.8	119
Iron	31	27.7	117
Lead	14.6	6.5	42
Manganese	0.83	0.54	2.7
Nickel	1.23	0.99	4.4
Selenium	.091	.061	0.30
Silver	.038	.029	0.13
Vanadium	.082	.076	0.32
Zinc	11.2	15.4	53
Aluminium	13.9	7.5	43
Indium	.004	.0039	0.020
Scandium	.003	.0027	0.010
Sodium	1.0	3.5	9.0

TABLE 4   Wet deposition ($\mu g.cm^{-2}$)[1]
August 1980–January 1981, and the projected annual deposition rate

Element	Aug–Oct	Nov–Jan	Projected annual rate ($\mu g.cm^{-2}.yr^{-1}$)	Rain-soluble %[2]
Antimony	0.072	0.14	0.42	58
Arsenic	0.39	∿0.3	1.30	∿40
Cadmium	0.26	–	1.04	∿80
Chromium	∿0.2	∿0.3	0.84	∿3
Cobalt	0.012	.006	.036	66
Copper	36	46	164	96
Iron	9.9	11	42	27
Lead	6.8	6.4	26	93
Manganese	0.26	0.91	2.3	70
Nickel	1.0	1.2	4.4	81
Selenium	0.094	0.08	0.35	67
Silver	–	0.03	0.12	∿60
Vanadium	0.086	0.1	0.37	49
Zinc	29	31	120	97
Aluminium	11.0	8.8	40	32
Indium	0.004	–	0.016	60
Scandium	–	0.001	0.004	<30
Sodium	21	85	210	99
Rainfall (mm) [3]	302	180	964	

NOTES

(1) Measured quantities are concentrations in rain (μg element per litre rain) for separate rain-soluble and rain-insoluble fractions. These are converted to depositions ($\mu g.cm^{-2}$) by multiplication by the rainfalls for the sampling periods (measured separately).

(2) Percentage rain-soluble material in the projected annual deposition

(3) Measured at site

TABLE 5 Deposition velocities
August 1980 – January 1981

Element	Deposition velocity $(mm.sec^{-1})$		Ratio Dry/Wet
	Dry $(v_d)^{(1)}$	Wet $(v_w)^{(2)}$	
Antimony	4.7	3.3	1.4
Arsenic	1.9	1.3	1.5
Cadmium	–	4.3	–
Chromium	15	4.8	3.1
Cobalt	14	5.7	2.5
Copper	14	19	0.73
Iron	16	5.6	2.8
Lead	4.6	2.9	1.6
Manganese	4.3	3.6	1.2
Nickel	10	10	1.0
Selenium	1.0	1.2	0.86
Silver	2.3	2.1	1.1
Vanadium	4.6	5.3	0.86
Zinc	2.0	4.6	0.44
Aluminium	20	18	1.1
Indium	7.9	6.3	1.3
Scandium	11	3.8	2.5
Sodium	1.6	37	0.043

$$(1) \quad v_d \ (mm.sec^{-1}) = \frac{\text{dry deposition rate } (\mu g.m^{-2}.sec^{-1})}{\text{concentration in air } (\mu g.m^{-3})} \times 10^3$$

$$(2) \quad v_w \ (mm.sec^{-1}) = \frac{\text{wet deposition rate } (\mu g.m^{-2}.sec^{-1})}{\text{concentration in air } (\mu g.m^{-3})} \times 10^3$$

TABLE 6 Correlations

Pairs of parameters	Correlation Coefficient	Degrees of Freedom	Significance Level
Wet and dry deposition[1]	0.70	14	> 99%
Air conc. and wet deposition[1]	0.76	15	> 99.9%
Air conc. and dry deposition[1]	0.57	14	> 95%
Dry/wet ratio and solubility	−0.74	15	> 99%
Dry dep.vel. and solubility	−0.48	15	> 90%
Dry dep.vel. and dry/wet ratio	0.45	15	> 90%

(1) Excluding Na results

1.0 to 20 mm sec^{-1} for v_d and 1.2 to 37 mm sec^{-1} for v_w.

The dry to wet ratios are also shown in the table, and demonstrate that the two mechanisms of deposition are roughly similar in magnitude in this urban polluted area. The chief exception is Na, with a ratio of only 0.043. The elements Cr, Co, Fe and Sc have ratios of >2.

There is a significant correlation between the dry and wet deposition rates, as shown in Fig 2, where Na is a prominent exception.

It is also interesting to compare the rain soluble fraction with the dry to wet ratio, shown in Fig 3, where there is a significant anti-correlation. There is a striking difference between the ratio for Na and for other elements with high solubility, such as Cu, Pb and Zn with ratios of 0.44 to 1.6. It seems likely that even in the most central parts of England, which may be about 150 km from the sea, most atmospheric Na comes from sea spray. In contrast the other elements in this study are likely to come from much closer sources, and for these, the dry deposition process is more favoured, even for very soluble elements.

Other correlations examined are shown in Table 6.

The concentrations in air correlate with both wet and dry deposition rates, but the correlation with the wet is better. A possible reason for this is the dry deposition process tends to apply to larger particles than the wet. Standard aerosol samplers tend to work with reduced efficiency for particles above about 5 μm aerodynamic diameter in ambient outside conditions[5], and hence the wet deposition material is more closely related to that collected by the aerosol sampler.

It has been shown elsewhere[2,6] that the v_d values are related to the mass median aerodynamic diameter (MMAD) of the airborne particle size distribution. The relationship is roughly linear, with a v_d of 10 mm sec^{-1} being equivalent to a MMAD of 2 μm. This suggests that, in the present measurements the elements, Cr, Co, Cu, Fe, Ni, Al and Sc are associated with particles having size distributions with an MMAD of 2 μm or more, and the elements Sb, As, Pb, Mn, Se, Ag, V, Zn and Na with a MMAD of 1 μm or less. It is considered[7] that particles having distributions with MMAD less than 1 μm are likely to be deposited in the lung after inhalation, whereas particles having distributions with MMAD greater than 2 μm are more likely to be trapped in the nose and throat.

The values of v_d show a moderate correlation with the dry-to-wet ratios, and a moderate anti-correlation with the rain-soluble fraction. These can perhaps be explained in terms of particle size, known to correlate with v_d. It suggests that elements which are associated with larger particles tend to have a lower rain soluble fraction than those associated with smaller particles.

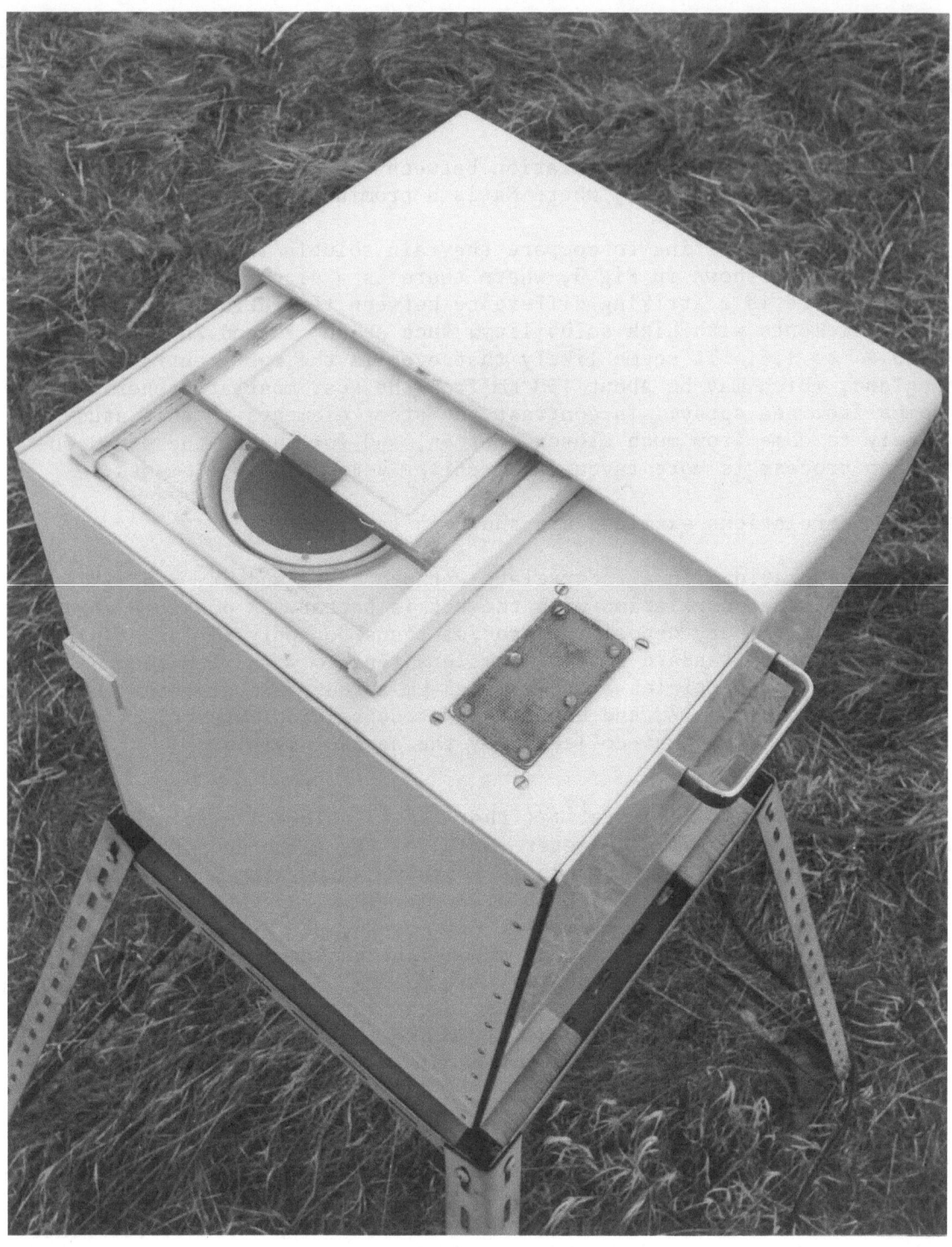

Figure 1 : Automatic wet and dry deposition collector.

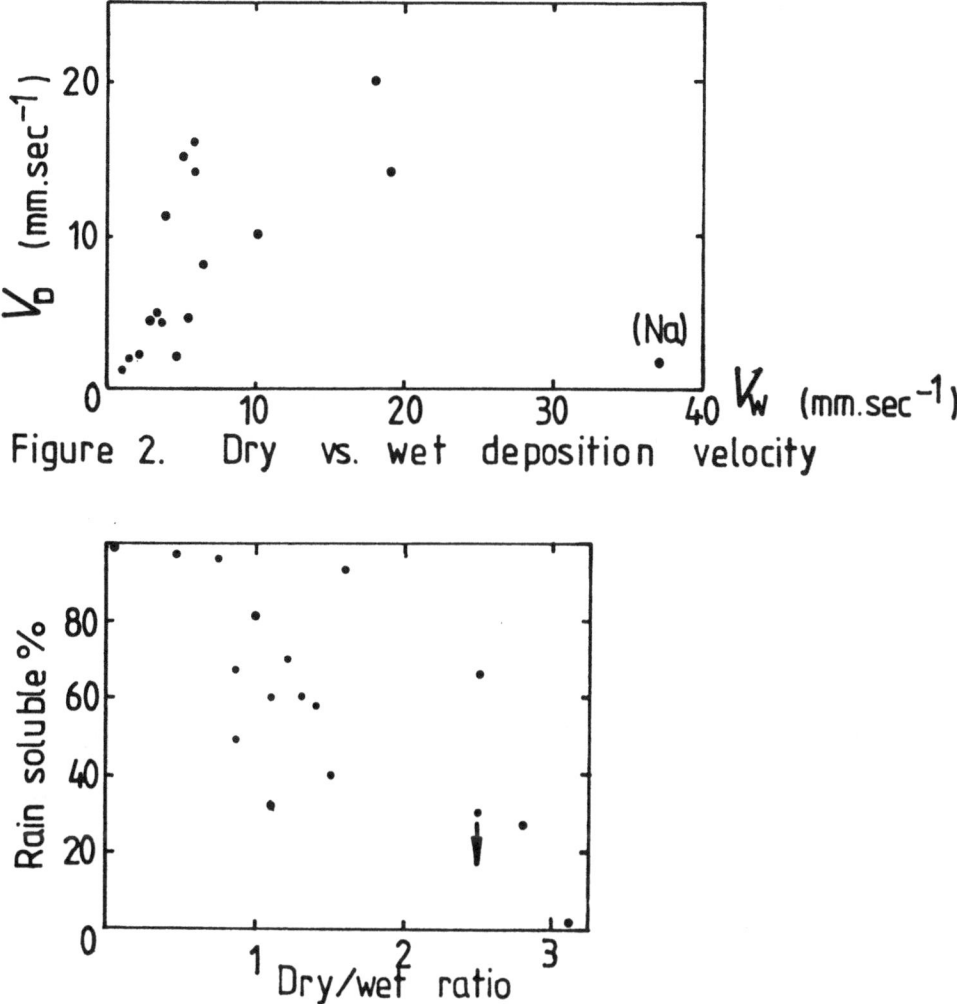

Figure 2.   Dry   vs.   wet   deposition   velocity

Figure 3.   Rain   soluble   fraction   vs.
            dry/wet   ratio   in   deposition

5.   CONCLUSIONS

(i)   The present measurements were performed in an area
subject to considerable atmospheric pollution with
heavy metals. The conclusions based on these measurements do not necessarily apply to other less polluted
areas.

(ii)   For most elements observed, the dry and wet deposition
rates are roughly similar, and they correlate well with
each other. The principal exception to this is sodium,
for which the dry deposition is much less than the wet.
Most of the sodium probably comes from distant sea spray.

(iii) There are indications that many elements which have
greater rain soluble fractions in wet deposition are
associated with smaller particles in the outside aerosol.
Most of these are also elements which may be emitted
to the air in industrial high temperature operations,
e.g. smelting, possibly initially in vapour form, and
which subsequently condense to form sub-micron sized
particles.

REFERENCES

1.   Salmon, L. "Instrumental neutron activation analysis in
studies of trace elements". AERE-R7859 (1975).
2.   Cawse, P.A. "A survey of trace elements in the UK (1972-73)".
AERE-R7669 (1974).
3.   Cawse, P.A. "Deposition of trace elements from the atmosphere in
the UK". Inorganic Pollution and Agriculture. MAFF Reference
Book 326, HMSO (1980).
4.   Chamberlain, A.C. "Aspects of the deposition of radioactive and
other gases and particles". Int. J. Air Poll. 3 (1960) 63-88.
5.   Pattenden, N.J., and Wiffin, R.D. "The particle size dependence
of the collection efficiency of an environmental aerosol sampler".
Atmos. Envir. 11 (1977) 677 681.
6.   Pattenden, N.J. "Atmospheric concentrations and deposition rates
of some trace elements measured in the Swansea/Neath/Port Talbot
area". AERE-R7729 (1974).
7.   Bates, D.V., Fish, B.R., Hatch, T.F., Mercer, T.T. and Morrow, P.E.
(Task Group on Lung Dynamics). "Deposition and retention models
for internal dosimetry of the human respiratory tract".
Health Physics 12 (1966) 173.

Deposition on Plants and Vegetation

CONCENTRATION OF AEROSOL CONSTITUENTS ABOVE AND BENEATH A BEECH AND A SPRUCE FOREST CANOPY

Gode Gravenhorst
Lab.de Glaciologie et Geophys.del'Environ.,CNRS
2 Rue Très Cloitres, F-38031 Grenoble, France

Klaus Dieter Höfken
Institute for Meteorology and Geophysics
Frankfurt/Main

ABSTRACT

Concentrations of atmospheric aerosol constituents were measured above and beneath the canopies of a beech- and a spruce forest in order to determine the filtering effect of a closed stand of trees. Aerosol concentrations beneath the canopies were significantly lower than those above. In the beech forest, this difference was more pronounced during the leafy period than during the leafless time. The filtering efficiency showed a strong dependence on the size of the aerosol particles with a minimum for particles in the medium size range of about 0.5 µm radius. The results are compared with dry deposition rates obtained by chemical analyses of precipitation collected above and beneath the tree canopies.

INTRODUCTION

Nearly one third of Germany's surface area is covered with forest. A considerable influence of this large surface on deposition and concentration of the atmospheric aerosol is expected (Knabe 1977, Ulrich et al. 1979). But hitherto there is only qualitative information on the scavenging of atmospheric aerosol particles by forests with regard to concentration (Blum 1965, Flemming 1972, Wentzel 1967) although the importance of forests and trees in reducing immissions is known. Therefore, this investigation was carried out to give quantitative information on the influence of two forest types on size, concentration and composition of the atmospheric aerosol.

H.-W. Georgii and J. Pankrath (eds.), Deposition of Atmospheric Pollutants, 187–190.
Copyright © 1982 by D. Reidel Publishing Company.

METHODS

During the year 1980 aerosol samples were taken simultaneously outside, above and beneath two forest canopies. The experimental site was located in the Solling forest, a rural area of extended woodlands in Central Germany, 500 m above sea-level. The two forests consist of about 25-30 m high trees of beech (fagus silvatica, ca.130 years) and spruce (picea abies, ca. 95 years) (more detailed in Ellenberg 1971).

Most aerosol samples were taken with 8 identical cascade impactors, constructed for this very purpose. In order to guarantee isokinetic sampling conditions, the impactors (4 stages plus back-up filter) were connected with wind vanes that moved the impactors with their inlets automatically into the wind direction. Moreover, inlets with different diameters suitable to the prevalent wind speeds were used. The aerosol was fractionated into the size ranges
>1.7 μm; > 0.9 μm; > 0.4 μm; > 0.2 μm; < 0.2 μm.

The samples were analyzed for the heavy metals Mn, Fe, Cd, Pb (atomic absorption) and the soluble ions NH_4^+ (colorimetry, indophenolblue method), Cl^-, NO_3^-, $SO_4^=$ (ion chromatography). A more detailed description of the methods is given by Höfken et al. 1981.

RESULTS

The above forest concentrations of the elements and ions that were investigated show a difference of several orders of magnitude (Tab. 1) with values that are characteristic for a rural area in Central Europe.

	MMD μm	concentration above forest ng/m³(100%)	concentration beneath the canopy (in %) beech	spruce
Cd	.26	1.3	66	59
$SO_4^=$	.63	8400	71	68
NH_4^+	.70	3000	73	72
Pb	.78	100	71	67
Mn	1.1	24	75	77
Cl^-	1.5	130	66	62
Fe	2.2	380	63	61
NO_3^-	2.4	1600	63	60

Table 1:   (summer (May - Oct) 1980)

Beneath the forest canopies the concentration was significantly lower during the whole period of investigation. During the summer months a decrease of concentration of about 25 to 40% was found within the beech and the spruce forest. The different surface structures of the two forest types showed a measureable influence on the concentration in samples collected above and beneath the forests. A significantly higher filtering efficiency of the spruce canopy is reflected in concentrations that are lower in the spruce than in the beech forest (Tab. 1). During the winter months, when the branches of the beech were leafless, the concentration differences in measurements outside and within the forest were only about half of the summer values.

The scavenging of aerosols from the atmosphere is a consequence of the deposition of the particles on the branches of the trees. Therefore, a dependence of the filtering efficiency on the diameter of the particles should be expected as well as the dependence of the deposition velocity on the particle-diameter.

In contrast to earlier investigations, when only few data were available (Höfken and Gravenhorst 1980), an influence of the particle-diameter on the filtering efficiency could now be confirmed. A minimum effect (highest concentration within the forests) was found for mass mean diameters (MMD) in the range of about $0.7-1\mu m$ (Tab. 1). The different scavenging efficiency of diffusion, sedimentation and impaction is not only reflected in a minimum of the deposition velocity within this size range (Sehmel and Hodgson 1974), but can also be detected in filtering measurements.

This work is part of the research project 104 02 600 of the Umweltbundesamt (Berlin), principal investigator Prof. H.-W. Georgii. We thank Prof. R. Mayer and Prof. B. Ulrich (Göttingen) for their helpful support.

REFERENCES

Blum, W. (1965) Forst- und Holzwirt 20, 211-215

Ellenberg, H. (1971) Integrated Experimental Ecology, Berlin

Flemming, G. (1972) Arch. Naturschutz und Landschaftsforschung 12, 177-188

Höfken, K.D. and Gravenhorst, G. (1980) Proceedings,

Aerosols in Science, Medicine and Technology, 8 Conference GAF, Schmallenberg, 37-42

Höfken, K.D., Georgii, H.-W. and Gravenhorst, G. (1981), Berichte des Institutes für Meteorologie und Geophysik, Universität Frankfurt, 46

Knabe, W. (1977) Proceedings, 4. Clean Air Congress, Tokyo, 952-957

Sehmel, G.A. and Hodgson, W.H. (1974) CONF 740921, 1976, 399-422

Ulrich, B., Mayer, R. and Khanna, P.K. (1979) Schriften aus der Forstl. Fakultät der Univ. Göttingen und der Niedersächs. Forstl. Versuchsanstalt, 58

Wentzel, K.F. (1967) Angew. Bot. 40, 1-11

DEPOSITION OF ATMOSPHERIC AEROSOL PARTICLES TO BEECH- AND SPRUCE FOREST

Klaus Dieter Höfken
Institute for Meteorology and Geophysics
Frankfurt/Main

Gode Gravenhorst
Laboratoire de Glaciologie, CNRS, Grenoble

ABSTRACT

Trace constituents are removed from the atmosphere either by incorporation into rain or by deposition onto surfaces like soil, vegetation, waters. Dry deposition onto natural surfaces can hardly be represented by mathematical models or by artificial surfaces, for realistic approximations of the complex boundary (e.g. vegetation/atmosphere) is not yet possible. Therefore, it was attempted to determine the deposition of atmospheric trace constituents to a natural surface in a direct way. A beech- und a spruce forest were used as a natural surface. The amount of matter dry deposited on the vegetation surface is derived by means of the difference of concentrations in precipitation above and beneath the canopy area. Leaching and absorption of certain substances by leaves and twigs are taken into consideration. Dry deposition velocities are determined from the transfer rates (atmosphere to vegetation) of the aerosol compounds and their simultaneously measured air concentration. They will be discussed with results of other investigations.

INTRODUCTION

Deposition is the most important process to reduce the atmospheric concentration of air pollutants. Unfortunately, our knowledge of the mechanism of deposition and of the transport of trace substances to natural surfaces like water, soil, vegetation is still very insufficient. There exists the need to understand the process of removal of trace substances from the atmosphere better, because con-

191

H.-W. Georgii and J. Pankrath (eds.), Deposition of Atmospheric Pollutants, 191–194.
Copyright © 1982 by D. Reidel Publishing Company.

nections between deposition of pollutants and damage to
ecosystems, especially forests, become obvious (BRAEKKE
1976, KNABE 1976, ULRICH et al. 1979).
This is the background for the present investigation which
deals with the deposition of aerosols to two forest ecosystems and gives a first survey of deposition velocities
of several atmopheric aerosol constituents upon a forest
canopy.

METHODS

At the site of the Solling forest where our investigations were performed, aerosol and precipitation samples
were taken during 1980. Dry and wet deposition rates were calculated by means of precipitation chemistry. A total of 20
precipitation funnels (25 cm dia., polyethylene) were
placed above and outside the beech and spruce forest respect.
to collect wet deposition and canopy throughfall of each
single rain event.
To assess the leaching of ions out of the plant substance
by rain, some beech and spruce twigs were washed with rain
water of known chemical composition. Some of the twigs had
been protected against dry deposition with plastic bags
that were ventilated with aerosol-free air for a period of
a week. The amount of lead that was absorbed on the twigs
and did not enter the throughfall could be calculated from
litterfall analyses by Heinrichs & Mayer (1980).

The samples were analyzed for the soluble ions Cl^-, NO_3^- ,
SO_4^- (ion-chromatography), NH_4^+ (colorimetry, indophenol-
blue method) and the soluble and insoluble fraction of the
heavy metals Mn, Fe, Cd, Pb (atomic absorption). Those
precipitation samples that could not be analyzed immediately, were stored deep frozen to keep the initial chemical
composition of the water samples.

RESULTS

Tab. 1 shows that the concentration of the ions and
elements in the throughfall was significantly (1.5 to
41.2 times) higher than in the precipitation above the
forest.

Tab. 1:   Concentration in precipitation (May-Oct 1980)

	NH_4	Cl	Pb	Fe	Cd	NO_3	SO_4	M_n
above forest (µg/l)	1010	970	15	115	3.6	2830	4100	22
beneath beech +	1.7	1.5	1.6	2.5	3.8	1.8	2.9	26.5
beneath spruce +	3.3	3.9	3.9	6.7	6.7	7.4	13.4	41.2

+   ratio of the concentration in the throughfall to the above-forest concentration

The values are in good agreement with the findings by Ulrich et al. (1979), Mayer (1981) who took precipitation samples over a period of several years. The concentration enrichment is due to the throughfall containing aerosol particles that were deposited on the branches during the dry period. Only in the case of Mn a considerable leaching was observed , so that the enrichment of Mn is due to both dry deposition and leaching. In Tab. 2 a compilation of deposition fluxes to the forest canopy is given:

Tab. 2:   Deposition rates in mg/m^2 month

components	wet deposition		dry deposition rate			
			beech		spruce	
	W	S	W	S	W	S
M_n	2	2	< 8	< 6	< 9	< 18
Fe	13	10	11	10	26	23
Cd	0.04	0.03	0.09	0.06	0.12	0.08
Pb	1.3	1.2	1.1	2.3	4.3	3.3
NH_4^+	130	84	61	28	120	61
Cl^-	160	80	6	13	90	67
NO_3^-	360	230	97	90	1200	670
$SO_4^=$	660	340	350	440	2600	1400

(W: Feb-Apr 1980, S: May-Oct 1980)

With these values and the corresponding concentrations in the atmospheric aerosol measured over the same period dry deposition velocities can be calculated (Tab. 3):

Tab. 3:   Deposition velocities for beech (summer 1980)

	Cd	SO_4	NH_4	Pb	M_n	Cl	Fe	NO_3
deposition velocity (cm/s)	1.8	1.1	1.0	0.9	0.7	1.0	1.0	1.3
mass mean diameter (µm)	0.26	0.63	0.70	0.78	1.1	1.5	2.2	2.4

The different surface structure of the trees is reflected
in their different influence on dry deposition. The deposition rate to spruce forest is higher than to beech
forest (Tab. 2). In winter, when the beech canopy is leafless, the difference between deposition rates to spruce
and beech is more pronounced and the beech deposition
rates are lowest (Tab. 2).

The deposition velocities presented in this paper reveal
relatively high values when compared with a compilation
of laboratory measurements (Sehmel 1980). A dependence
between particle size and deposition velocity was found
with minimum values in the range of 1 μm MMD (Tab. 3).
This minimum and the different extent of the elements,
deposition is well confirmed by aerosol and filtering
measurements.

This work is part of the research project 104 02 600 of
the Umweltbundesamt (Berlin), principal investigator
Prof. H.-W. Georgii. We thank Dr. R. Mayer and Prof.
B. Ulrich (Göttingen) for their helpful support.

REFERENCES

Braekke, F. H.:   1976, fagrapport FRG/76,
                  SNSF-prosjektet, Oslo-Aas
Heinrichs, H. and Mayer, R.: 1980, Journal of Environ-
                  mental Quality 6, 111-118
Knabe, W.: 1976, AMBIO 5, 213-218
Mayer, R.: 1981, Göttinger Bodenkundl. Berichte 70
Müller, K. P. and Gravenhorst, G.: this publication
Sehmel, G.A.: 1980, Atmospheric Environment 14,983-1011
Ulrich, B., Mayer, R. and Khanna, P. K.: 1979,
          Schriften aus der Forstlichen Fakultät der
          Universität Göttingen und der Niedersächsischen
          Forstlichen Versuchsanstalt, 58

CALCULATION OF DEPOSITION RATES FROM THE FLUX BALANCE AND ECOLOGICAL EFFECTS OF ATMOSHPERIC DEPOSITION UPON FOREST ECOSYSTEMS

Robert Mayer and Bernhard Ulrich
Institut für Bodenkunde und Waldernährung der
Universität Göttingen, Büsgenweg 2, D-3400 Göttingen

ABSTRACT

Total wet and dry deposition of main and trace elements is calculated from the flux balance for precipitation fluxes of forest canopies for which precipitation fluxes above and below canopy as well as the flux coupled with litterfall has been measured. Retention and leaching of substances are estimated for groups of elements individually showing similar chemical and physiological behavior. It is shown that dry deposition may be very important and may even exceed wet deposition to forests in rural areas. Dry deposition rate in the same area depends clearly upon forest type, i. e. upon the surface quality. Ecological consequences of atmospheric deposition to forests are discussed.

MODE OF CALCULATION

Total atmospheric deposition to a surface is the sum of two components, wet and dry deposition. While it is comparatively easy to measure wet deposition, the assessment of dry deposition to a surface covered with vegetation can not be done in a simple, direct way.

Atmospheric substances deposited to a forest canopy will partly be dissolved in the rain before reaching the canopy, or washed off and be dissolved after deposition on the canopy surface. Therefore, precipitation collected below a forest canopy will normally contain a larger element load than precipitation collected above the canopy. In addition to the dissolved fraction of dry deposition, precipitation below the canopy may contain substances leached from vegetation surfaces, originating from internal turnover (i.e., taken up from soil by roots).
The element load of precipitation below a vegetation canopy, expressed in the dimensions of a flow rate, can be written

195

H.-W. Georgii and J. Pankrath (eds.), Deposition of Atmospheric Pollutants, 195–200.
Copyright © 1982 by D. Reidel Publishing Company.

in the form of a balance equation:

$$P_c = D_d + D_w - R_c + L_c \qquad\qquad (1)$$

where P_c is the element flux below the canopy, D_d is dry deposition, D_w is wet deposition to the canopy, R_c is the retention of atmospheric substances within the canopy, L_c is canopy leaching.

The only flux on the r.h.s. of equation (1) that can be measured directly is wet deposition, D_w. Also P_c can be easily measured. For many elements at least two of the remaining fluxes can be measured indirectly or estimated with reasonable precision so that the third flux can be calculated. For each element the best way for the assessment of all fluxes contributiong to the element load of precipitation reaching the ground must be discussed individually according to its individual chemical behavior and its changing role in physiological processes of forest vegetation.

According to their way of assessment, the elements can be combined in the following groupings:

(1) The first group includes the elements S, Na, K, Mg, Ca, Mn, Fe, and Zn (for details see Ulrich et al. 1979). These elements show predominantly high solubility on the forest canopy, at least under acid rain conditions. Therefore retention due to physico-chemical processes becomes negligible, while uptake due to metabolic processes can be ruled out because the elements contribute relatively little to the element content of organic materials (S, Na, Ca, Zn), or are leached from the canopy in considerable quantities (K, Mg, Zn, Mn) during the vegetation period. During the dormant season, i.e. the winter months, when beech has lost its leaves, leaching of metabolites from the canopy becomes negligible. This fact allows to estimate dry deposition from the balance equation (1). Dry deposition during the vegetation period is then derived from the wet/dry deposition ratio estimated for the winter period.

(2) The second group of elements includes Al and the heavy metals Cr, Co, Ni, Cu, Cd, and Pb (for details see Mayer 1981). For the conditions of a forest vegetation it can be assumed that leaching of these elements which are considered relatively immobile is negligible. It has been established in numerous experiments reported in literature that uptake of these metals via roots is restrained compared to that of nutrient elements. Measurements of the heavy metal concentration in the roots of different size

classes in the forests under investigation (Mayer und Heinrichs 1981) pointed into the same direction. Under these
conditions a rough estimate of the maximum uptake rate can
be made. This estimate is small compared to total annual uptake into aboveground biomass which therefore must be sup-
plied predominantly from external sources, i.e. from atmospheric deposition. Since total annual uptake can be determined experimentally the very rough estimate of root uptake
can be used to calculate quite precisely the rate of retention which then can be substituted in equation (1) to be
solved for the dry deposition rate.

At this point it is necessary to discuss the reliability of
these estimates. This will be done for two elements, one
from the first group, sulfur, and one from the second group,
lead. Table 1 shows measured data for the sulfur balance of
a beech ecosystem in the Solling, averaged over a period of
8 years (1968-1976). It can be seen that a large portion

Table 1. Sulfur fluxes in a beech ecosystem
(Solling B1)   1968-1976 average

(1) Measured rates
   - Wet deposition (soluble fraction
     found in bulk samplers)   23.8 $kgS.ha^{-1}.a^{-1}$
   - Wet deposition under canopy
     (stemflow included)   53.1 $kgS.ha^{-1}.a^{-1}$
   - Total uptake of S into above-
     ground biomass per year (from
     Meiwes 1978)   10.1 $kgS.ha^{-1}.a^{-1}$
   - Dry deposition during the dormant
     season (Nov-April) assuming lea-
     ching and retention to be zero   13.9 $kgS.ha^{-1}$

(2) Estimates
   - Dry deposition during vegetation
     period (May-October)   12.0 $kgS.ha^{-1}$

(3) Total deposition calculated from
    flux balance (1)   49.7 $kgS.ha^{-1}.a^{-1}$

---

of total deposition can be measured directly, namely wet deposition and dry deposition during the winter months. Dry
deposition estimates for the vegetation period can be considered to be quite safe since total uptake into the aboveground biomass is found to be small, a fact which would not
allow retention to play a very important role.

Table 2 shows measured data for the lead balance of the same
beech ecosystem, averaged over the period of 5 years (1974-
1979). Since Pb is considered to be very immobile within the

plant, which means that it is largely excluded from root uptake and held back in the rhizosphere or root surface, leaching can be assumed to be negligible. From measured data

---

Table 2.   Lead fluxes in a beech ecosystem
           (Solling B1)   1974-1979 averages

(1) Measured rates

   -Wet deposition (soluble fraction
    found in bulk samplers)   285 $gPb.ha^{-1}.a^{-1}$
   -Wet deposition under canopy
    (stemflow included)   302 $gPb.ha^{-1}.a^{-1}$
   -Total uptake of Pb into the
    aboveground biomass per year   169 $gPb.ha^{-1}.a^{-1}$

(2) Estimate

   -Maximum root uptake rate   34 $gPb.ha^{-1}.a^{-1}$

(3) Total deposition calculated   437 $gPb.ha^{-1}.a^{-1}$
    from flux balance (1)

---

of transpiration rate and lead concentration in soil solution the estimate for a maximum root uptake rate proves to
be very small compared to total uptake into the aboveground
biomass. This makes the estimate for the retention rate
quite safe.

RESULTS

The results of our measurements and calculations of deposition rates are given in table 3. The data show that total deposition of atmospheric substances to a forest covered surface can exceed wet deposition rates collected in a
bulk rain gauge (a so-called wet plus dry sampler) considerably. It can be seen furtheron that dry deposition to a
spruce canopy exceeds that to a beech canopy in the same
area as a consequence of differences in the absorption
characteristics of the canopies. The deposition excess of
spruce compared to beech is mainly occuring during the winter months when beech has lost its leaves and has therefore
a reduced surface area compared to the evergreen spruce
canopy.

The important contribution of dry deposition in the study
area is assumed to be due to the geographic situation. The
Solling is characterized by high precipitation, frequent
fog with predominant western and southwestern winds. This
is also the direction in which large industrial complexes
are found, namely the area around Kassel (50 km) and, more
important, the Ruhr industrial zone (130 km). Coming from

Table 3. Annual deposition rates for a beech (B) and
a spruce (S) stand in the Solling

Element		Wet deposition	Dry deposition	Total deposition
		$kg.ha^{-1}.a^{-1}$		
S	B	23.8	25.9	49.7
	S	23.8	49.8	73.6
Cl	B	16.9	13.2	30.1
	S	16.9	19.0	35.9
Na	B	7.7	4.7	12.4
	S	7.7	8.3	16.0
K	B	3.3	8.8	12.1
	S	3.3	18.2	22.5
Mg	B	1.9	2.0	3.9
	S	1.9	2.9	4.8
Ca	B	10.4	10.8	21.2
	S	10.4	16.9	27.3
Al	B	1.1	1.1	2.2
	S	1.1	1.8	2.9
Cr	B	0.014	0.135	0.149
	S	0.014	0.152	0.166
Mn	B	0.26	1.5	1.76
	S	0.26	4.9	5.2
Fe	B	0.88	0.89	1.77
	S	0.88	1.24	2.12
Co	B	0.014	0.001	0.015
	S	0.014	0.002	0.017
Ni	B	0.027	0.096	0.123
	S	0.027	0.113	0.140
Cu	B	0.236	0.234	0.470
	S	0.236	0.420	0.659
Zn	B	1.377	0.255	1.632
	S	1.377	0.355	1.732
Cd	B	0.016	0.0	0.016
	S	0.016	0.004	0.020
Pb	B	0.285	0.152	0.437
	S	0.285	0.448	0.732

1968-1976 average for S, Cl, Na, K, Mg, Ca, Al, Fe, Mn

1974-1979 average for Cr, Co, Ni, Cu, Zn, Cd, Pb

these areas, the Solling forms the first high altitude pla-
teau.

ECOLOGICAL CONSEQUENCES AND CONCLUSIONS

There is a number of elements including K, Mg, and Ca,which
can be considered as beneficial for the forest ecosystems,
taking into account that these are nutrient elements the
introduction of which into the system does not lead to any
adverse chemical reactions. Other elements like S, but also
N, are nutrients as well but deposition of these is connected with the simultaneous deposition or generation of acidity, or hydrogen ions, in the precipitation reaching the
surface. This has several effects, the most important of
which is the enhanced weathering of soil minerals with subsequent mobilization of toxic Al and heavy metals in the
root zone of forest vegetation. It is most likely that in
acid, low-buffered soils these processes give rise to root
damage and a destabilization of the forest ecosystem.

The same can take place as a consequence of heavy metal deposition. These elements, Pb, Cd, Cu, and Zn in the first
line, may have adverse effects when retained in the canopy
by damaging microorganisms on the leaf surfaces, epiphytes
or the leaves themselves. When heavy metals reach soil with
precipitation or together with litterfall they are, in most
cases, accumulated in the humus layer covering the mineral
soil. This leads to the buildup of a pool of toxic elements
that can be readily mobilized when organic matter is decomposed.

As a conclusion I would like to stress the following points:
The assessment of deposition rates in the described way can
not substitute for exactly performed measurements according
to meteorological methods. Our method reveals relatively
little information on deposition processes but it leads to
long-term avarages of deposition rates urgently needed by
ecologists, soil scientists, geochemists and enviromental
protection engineers dealing with element cycling and
land pollution problems. For the meteorologist, on the
other hand, our data may furnish the framework in the space
and time scale in which their investigations on deposition
processes and deposition rates can be fitted.

REFERENCES

Mayer, R.:1981, Gött.Bodenk.Ber.70, pp.1-292
Mayer, R., und H.Heinrichs: 1981, Z.Pflanzenernähr.Bodenk,
    144, pp. 637-646
Meiwes, K.J.: 1978, Gött.Bodenk.Ber. 60, pp.1-108
Ulrich,B.,R.Mayer, P.K.Khanna:1979,Schr.Forstl.Fak.Univ.
    Göttingen u.Nds.Forst.Vers.Anst. 58, pp.1-291

EFFECTS OF ATMOSPHERIC POLLUTANTS ON MATERIALS; RESEARCH NEEDS

R.W. Lanting
TNO Research Institute for Environmental Hygiene
Delft, the Netherlands

ABSTRACT

When pollutants are removed from the atmosphere by deposition another
environmental compartment will be effected.
This presentation will outline the problems encountered when the effects
on man-made materials have to be quantified.

1. INTRODUCTION

This colloquium is intended to discuss the removal rates of pollutants
from the atmosphere and the mechanisms governing these processes. This
paper will illustrate the problems encountered when one's scientific
knowledge is to be applied to practical goals like quantifying the effects of air pollution in support of decision making on administrative
level. This will be demonstrated by the example of atmospheric corrosion
of man-made materials.

2. THE PROBLEM OUTLINED

Deposition rates are very important as they have to be incorporated in
a quantitative way in dispersion models in order to be able to give
sensible predictions.
These models are becoming more and more popular as a tool for air
quality management and will eventually make quite a lot of air monitoring activities redundant. As a matter of fact the National Monitoring
Network in the Netherlands will be reduced by at least hundred stations
in the coming years.
However, removal of pollutants from the air compartment means the input
into another environmental compartment: soil, water, biosphere.
The impact of this input on these systems is widely unknown. Although
there is a wealth of air quality data, the data on the effects of deposition are still very fragmentary.
Moreover our knowledge on the relations between emission, dispersion

H.-W. Georgii and J. Pankrath (eds.), Deposition of Atmospheric Pollutants, 201–206.
Copyright © 1982 by D. Reidel Publishing Company.

and deposition shows many scientific gaps.

Good decision making in environmental protection should rely to a large extent on the quantification of the effects of pollutants.

As these effects are seldom known the benefit of a regulating measure cannot be predicted and can only be estimated by long-term monitoring programs before and after a regulation is set into force.

The air quality situation in the Netherlands is changing by the gradual switch over from low sulfur natural gas to oil and coal as the primary energy sources for power plants.

This will lead to higher SO_2 levels. Moreover waste incineration will locally give rise to increased concentration levels of hydrochloric acid. So an increasing agressiveness of the atmosphere is anticipated and the problem of material deterioration by atmospheric pollutants is again at issue.

Recent studies estimate the economic damage owing to material losses caused by air pollution between one and two percent of the gross national product [1]. Also from the point of view of conservation of resources this is a matter of concern. The uncertainty of these economic damage estimates is very high owing to the lack of reliable dose–effect functions. This has precluded their effective use as a guidance for decision making. Up to now the poorly understood deposition rates and the poorly characterized chemical and physical interactions make the development of quantitative dose–effect functions very difficult.

3. THE MECHANISM

The mechanisms causing material deterioration are very complex (Figure 1).

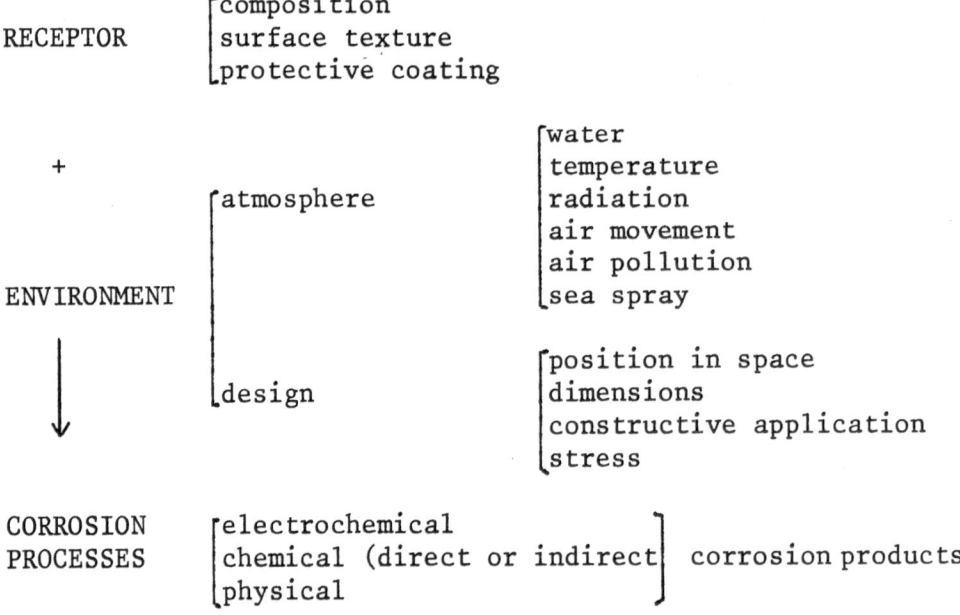

Figure 1. The system of atmospheric corrosion. After [1].

Deposition of air pollution on the receptor surface is only one of the
many factors causing material damage.
For many pollutants there is a threshold concentration below which no
effects occur.
The atmospheric factors can operate simultaneously or succesively. They
can reinforce or neutralize each other. Generally the actual corrosion
process is discontinuous, the presence of water being one of the most
essential parameters. The maritime environment in the Netherlands is
also a decisive factor.
The role of air pollution can be described by the processes of wet and
dry deposition. In Figure 2, the process of material attack by dry deposition in shown. The presence of water is an essential condition.

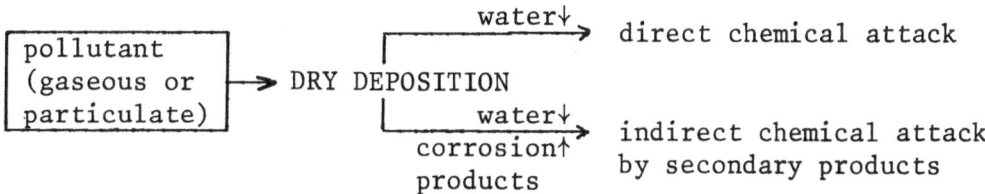

Figure 2. The role of air pollutants in the corrosion process.

The corrosion products themselves and the material characteristics
(porosity, catalytic properties) are important factors promoting the
formation of secondary more agressive compounds (e.g. H_2SO_4).
For some materials (copper, lead and zinc) the corrosion products act
as a protective coating preventing further chemical attack.
The role of wet deposition is rather complex.
According to the properties of the material the following processes
can be discerned:
- Wetting of the receptor surface. A precondition for the chemical attack to start.
- Dissolution of soluble corrosion products. This leads to two possibi-
  lities:
  1. when the corrosion products act as a protective layer the fresh
     surface is exposed to new attacks.
  2. when the corrosion products are an active factor in the process
     the chemical attack will be slowed down.
- Accumulation of dissolved corrosion products and chemical compounds
  present in the precipitation in crevices, pits and crannies where
  water is kept in place by capillary action.

For most materials the first two processes will counteract each other.
In practice for non-porous materials, one will notice an enhanced attack by atmospheric corrosion of those parts which are in the rain
shadow.
The third process is considered the most deleterious mechanism for
metal constructions.
So the role of air pollution in the atmospheric corrosion is governed
by wet and dry deposition, which cannot be seperated easily in their

effects.
Dry deposition of SO_2 on moist surfaces is a very dominant factor for
almost all materials. Conductivity of rain water and pH at values lower
than 4 correlate also well with the corrosion rate of metals.

4. CURRENT RESEARCH

Looking at the present knowledge we must conclude that most of the experimental results have only provided qualitative information that does
not meet the requirements of a real damage function. This is not surprising taking into account the tremendous problems involved in such
experiments. In general we can distinguish between two approaches.
Firstly, empirical studies looking for causitive relationships between
total damage and the variability in time and space of the atmospheric
parameters over a testing period. These studies aim at the abatement
of the cause and try to predict material behaviour in the real atmosphere.
Secondly, theoretical studies investigating the reaction mechanisms of
the corrosion process. The objective is abatement of the effect by material improvement.
In practice, the empirical approach is seriously hampered for the
following reasons:
- How is material damage to be quantified?
  For some materials quantification is by weight loss determination.
  However, loss of strength, soiling and other aesthetical aspects are
  hard to measure quantitively.
- How representative is a receptor for its real life application?
  For proper correlations with atmospheric factors, a material with a
  rapidly measurable corrosion rate is favoured. Such materials will
  usually not be applied without coating. Constructive applications like
  riveting, welding, etc. cannot be simulated easily in a field exposure
  experiment.
- How representative are the measurements of the environmental parameters?
  To establish a damage function or dose-effect relationship one needs
  to know the flux of pollutants to the receptor surface and the relationship between flux and air concentration. Deposition rates of
  pollutants for material surfaces under different climatological circumstances are not well known. In some empirical studies deposition
  rather than concentration is measured, although the deposition surface usually is not representative for the receptor under study. Moreover, there is no well-established relationship between deposition
  dose and   air concentration for such devices.
  Another problem lies in the fact that objects in which the materials
  are generally applied (buildings, constructions, cars, etc.) are subject to a microclimate which is quite distinct from the climatological parameters measured in the open field.
- How can causitive relations between atmospheric parameters and damage
  measured be established?
  Interactions between atmospheric parameters and their covariances

should be well understood in order to understand the meaningfulness
of correlations found. The results of theoretical studies might provide valuable information on this point. However, there is still
little exchange of knowledge between the two scientific approaches.
Not in the least for reasons of commercial protection and competition.

So, although there is a wealth of data from controlled exposure experiments, none of them can provide the right answer in terms of a real
dose-effect relationship [3]. Nevertheless, most causal relations are
well-established now.

5. NEW APPROACHES

With our increasing knowledge of deposition mechanisms and deposition
rates a new theoretical approach seems to be feasible, taking into account the already known causal relationships. If the deposition velocity of damage-causing pollutants on the different material surfaces can
be quantified, a mathematical model predicting the deposition rate and
the deposition dose can be developed and validated for a receptor surface under different meteorological conditions.
This would imply that deposition experiments which sofar mainly covered
the most important sinks (vegetation and soil) should also include man-
made materials in order to provide the basic data for the model. With
the present knowledge of the reaction kinetics of the corrosion processes and knowing the dose, we might come to an integrated model predicting effects. Such a model should be suitable to predict material
behaviour in a changing air quality situation and to estimate the benefit of an emission limit value or an air quality standard.
The model should be validated by field experiments, involving new sensitive techniques for the assessment of corrosion rates. Such techniques
can be found in electro-chemical methods, optical methods (light-scatte-
ring, interference patterns, image analyses, etc.) and new surface analysis techniques (PIXE, ESCA).
In such experiments it is very important to distinguish between wet and
dry deposition. The mechanism of wet deposition should be clarified.
If rain-out is the predominant mechanism, abatement strategies would be
of more than regional significance.
Considering the current application of materials in structures, priority should be given to the following materials:

uncoated metals	: galvanized steel, aluminum
coated materials	: painted carbon steel
building materials	: concrete, asbestos cement (as a possi-ble source of asbestos fibres owing to leaching)
materials of historical value	: calcareous stones (limestone, sandy limestone), combinations of brick and portland cement, documents in archives.

For the latter group of materials, the irreplaceability of these objects

is the main reason for concern rather than economic damage.
In these instances, solutions are generally sought in the abatement of
the effect by using conservation techniques.
Target pollutants associated with material deterioration are SO_2, NO_2
and their secondary products. In wet deposition chloride ions, hydrogen
ions and sulphate ions are of importance.
Incorporation of these materials and pollutants in deposition experiments would greatly contribute to a better understanding of material
damage owing to air pollution.

6. LITERATURE

[1] Arntzen, J.W.: Economic evaluation of damage to materials due to
                    air pollution, European Commission, 1980, report
                    EUR 6641 ENV.
[2] Sereda, P.J. : Effects of sulphur on building materials,
                    in: "Sulphur and its inorganic derivates in the
                    Canadian environment", Nat. Res. Council of Canada
                    NRCL no. 15015, pp. 354-426.
[3] Benarie, M. : Critical review of the available physico-chemical
                    material damage functions of air pollution, European
                    Commission, 1980, report EUR 6643 EN.

LIST OF PARTICIPANTS

Aheimer, G.	Kernforschungsanlage Jülich GmbH Institut f. Atmosphär. Chemie Postfach 913 5170 Jülich
Arnold, Dr. I.	Institut f. Meteorologie und Geophysik Universität Frankfurt Feldbergstr. 47 6000 Frankfurt/Main 1
Asman, Dr. W.	Institute for Meteorology and Oceano- graphy, 5 Princetonplein 3584 CC Utrecht, NL
Baltrusch, M.	Meteorologie Consult GmbH Dr. Rainer Schmitt Auf der Platt 47 6246 Glashütten
Beheng, Dr. K.	Institut für Meteorologie und Geophysik Universität Frankfurt Feldbergstr. 47 6000 Frankfurt/Main
Beilke, Dr., S.	Umweltbundesamt Pilotstation Frankfurt Feldbergstr. 45 6000 Frankfurt/Main 1
Beltz, N.	Referat f. Umweltschutz der Universität Frankfurt, Inst. f. Physikalische Chemie Robert-Mayerstr. 11 6000 Frankfurt/Main 1
Berresheim, H.	Referat f. Umweltschutz der Universität Frankfurt, Inst. f. Physikalische Chemie Robert-Mayerstr. 11 6000 Frankfurt/Main 1
Bingemer, H.	Institut f. Meteorologie und Geophysik Universität Frankfurt Feldbergstr. 47 6000 Frankfurt/Main 1
Bombelka,	Universität Marburg, Fachbereich Physik Renthof 5 3550 Marburg

Bräutigam, K.-R.   Kernforschungszentrum Karlsruhe
                          Abtlg. für Angew.Systemanalyse
                          Postfach 3640
                          7500 Karlsruhe 1

Brandtner, M.   Institut f. Meteorologie und Geophysik
                          Universität Frankfurt
                          Feldbergstr. 47
                          6000 Frankfurt/Main 1

Brosche, B.   Institut f. Meteorologie und Geophysik
                          Universität Frankfurt
                          Feldbergstr. 47
                          6000 Frankfurt/Main 1

Bürgermeister, S.   Institut f. Meteorologie und Geophysik
                          Universität Frankfurt
                          Feldbergstr. 47
                          6000 Frankfurt/Main 1

Davies,Dr. T. D.   School of Environmental Sciences
                          University of East Anglia
                          Norwich NR 47TJ, GB

Dimitri, Priv.Doz.   Hessische Forstliche Versuchsanstalt
    Dr.L.   Institut für Forstproduktion
                          Prof. Oelkers Str. 6
                          3510 Hann. Münden

Dongmann, Dr. G.   Kernforschungsanlage Jülich GmbH
                          Institut für Chemie ICH4
                          Postfach 913
                          5170 Jülich

Eisner, C.   Institut f. Meteorologie und Geophysik
                          Universität Frankfurt
                          Feldbergstr. 47
                          6000 Frankfurt/Main 1

Fricke, Dr. W.   Gesamtverband des Deutschen Steinkohle-
                          bergbaus
                          Friedrichstr. 1
                          4300 Essen

Garland, Dr. J. A.   Environmental and Medical Science Div.
                          AERE
                          Harwell, Didcot
                          Oxfordshire OX 11 ORA, GB

Georgii, Prof.,Dr.   Institut für Meteorologie und Geophysik
   H.-W.   Universität Frankfurt
                          Feldbergstr. 47
                          6000 Frankfurt/Main 1

Gravenhorst,Dr.,G.   Centre National de Recherche Scienti-
                          fique, Lab. de Glaciologie et Geophys.
                          de L'Environnement
                          2, rue Très Cloitres
                          38031 Grenoble-Cedex, F

Grosch, W.   Umweltbundesamt
                          Pilotstation Frankfurt
                          Feldbergstr. 45
                          6000 Frankfurt/Main 1

Haunold, W.   Institut f. Meteorologie und Geophysik
                          Universität Frankfurt
                          Feldbergstr. 47
                          6000 Frankfurt/Main 1

Helas, Dr.,G.   Max Planck Institut für Chemie
                          Postfach 3060
                          6500 Mainz

Herbert,Prof.Dr.F.   Institut für Meteorologie u. Geophysik
                          Universität Frankfurt
                          Feldbergstr. 47
                          6000 Frankfurt/Main 1

Herrmann,J.   Referat für Umweltschutz der
                          Universität Frankfurt
                          Inst. f. Physikalische Chemie
                          Robert-Mayerstr. 11
                          6000 Frankfurt/Main 1

Höfken, K.D.   Kernforschungsanlage Jülich GmbH
                          Institut f. Atmosph. Chemie
                          Postfach 913
                          5170 Jülich

Horvath, Dr.,L.   Institute for Atmospheric Physics
                          Meteorological Service of the Hungarian
                          P.O.B. No. 39
                          H-1675 Budapest

Jacobsen,Dr.L   Deutscher Wetterdienst
                          Frankfurterstr. 135
                          6050 Offenbach

Jaeschke, Dr. W.   Referat for Umweltschutz der Univ.
                          Frankfurt
                          Inst. f. Physikalische Chemie
                          Robert-Mayerstr. 11
                          6000 Frankfurt/Main 1

Kins, L.   Institut f. Meteorologie und Geophysik
                          Universität Frankfurt
                          Feldbergstr. 47
                          6000 Frankfurt/Main 1

Klockow,Prof.Dr.,D.   Chemisches Institut
                          Universität Dortmund
                          Postfach 500500
                          4600 Dortmund

Knabe,Dr.W.   Landesanstalt für Ökologie
                          Leibnizstr. 10
                          4350 Recklinghausen

Kramm, G.   Institut f. Meteorologie und Geophysik
                          Universität Frankfurt
                          Feldbergstr. 47
                          6000 Frankfurt/Main 1

Kuttler, Dr.W.   Universität Bochum
                          Geographisches Institut
                          Postfach 102148
                          4630 Bochum

Lanting,Dr.R.W.   Forschungsinstitut für Umwelthygiene
                          TNO, P.O. 214
                          2628 VK Delft, NL

Larner, Dr.,D.J.   D. Reidel Publishing Company
                          P.O. Box 17
                          3300 AA Dordrecht, NL

Löbel,Dr.,J.   VDI Kommission Reinhaltung der Luft
                          Graf Recke Str. 84
                          4000 Düsseldorf 1

Marggrander, E.   Ringstraße 25
                          6242 Kronberg/Taunus

Mayer, Prof.Dr.R.   Institut für Bodenkunde u. Waldernährung
                          der Universität Göttingen
                          Büsgenweg 2
                          3400 Göttingen

Müller, Dr. J.	Umweltbundesamt Pilotstation Frankfurt Feldbergstr. 45 6000 Frankfurt/Main 1
Müller, K.P.	Institut für Atmosph. Chemie ICH3 Kernforschungsanlage GmbH Postfach 913 5170 Jülich
Neuber, E.	Institut f. Meteorologie und Geophysik Universität Frankfurt Feldbergstr. 47 6000 Frankfurt/Main 1
Neumann-Hauf,G.	Kernforschungszentrum Karlsruhe Abteilung für Angew.Systemanalyse Postfach 3640 7500 Karlsruhe 1
Nguyen, V.D.	Institut für Chemie ICH4 Kernforschungsanlage GmbH Postfach 913 5170 Jülich
Nürnberg,Prof.Dr. H.-W.	Institut für Chemie ICH4 Kernforschungsanlage GmbH Postfach 913 5170 Jülich
Obeth, M.	Institut für Meteorologie u. Geophysik Universität Frankfurt Feldbergstr. 47 6000 Frankfurt/Main 1
Pankrath,Dr.J.	Umweltbundesamt Berlin Bismarckplatz 1 1000 Berlin 33
Pattenden,Dr.N.J.	Environmental Medical and Science Div. AERE Harwell, Oxfordshire OX 11 ORA, GB
Perseke, C.	Institut f. Meteorologie u. Geophysik Universität Frankfurt Feldbergstr. 47 6000 Frankfurt/Main 1

Pflock, H.   Institut f. Meteorologie und Geophysik
                            Universität Frankfurt
                            Feldbergstr. 47
                            6000 Frankfurt/Main 1

Rohbock, E.   Institut f. Meteorologie und Geophysik
                            Universität Frankfurt
                            Feldbergstr. 47
                            6000 Frankfurt/Main 1

Schmitt, G.   Institut f. Meteorologie und Geophysik
                            Universität Frankfurt
                            Feldbergstr. 47
                            6000 Frankfurt/Main 1

Schönwiese,Prof.   Institut f. Meteorologie und Geophysik
   Dr. C.-D.   Universität Frankfurt
                            Feldbergstr. 47
                            6000 Frankfurt/Main 1

Schreiber, B.   Meteorologie Consult GmbH
                            Dr. Rainer Schmitt
                            Auf der Platt 47
                            6246 Glashütten

Schurath,Prof.   Institut für Physikalische Chemie der
   Dr. U.   Universität Bonn
                            Wegelerstr. 12
                            5300 Bonn

Stein, D.   Institut f. Meteorologie u. Geophysik
                            Universität Frankfurt
                            Feldbergstr. 47
                            6000 Frankfurt/Main 1

Thorn, K.-F.   Institut f. Meteorologie und Geophysik
                            Universität Frankfurt
                            Feldbergstr. 47
                            6000 Frankfurt/Main 1

Valenta, Dr.P.   Institut für Chemie ICH4
                            Kernforschungsanlage Jülich GmbH
                            Postfach 913
                            5170 Jülich

Wagner, H.   Institut für Meteorologie u. Geophysik
                            Universität Frankfurt
                            Feldbergstr. 47
                            6000 Frankfurt/Main 1

Wallenwein, H.   Institut für Meteorologie u. Geophysik
                        Universität Frankfurt
                        Feldbergstr. 47
                        6000 Frankfurt/Main 1

Warneck,Prof.Dr.P.   Max-Planck-Institut für Chemie
                        Postfach 3060
                        6500 Mainz

Weber, Reg.Dir.E.   Bundesministerium des Innern
                        Rheinsdorferstr. 198
                        Postfach 170290
                        5300 Bonn 1

Wentzel, Dr. K.F.   Hessische Landesanstalt f. Umwelt
                        Mühlgasse 4
                        6200 Wiesbaden

Winkler, Dr.,P.   Deutscher Wetterdienst
                        Meteorologisches Observatorium
                        Frahmredderstr. 95
                        2000 Hamburg-Sasel

Zippel, N.   Institut für Meteorologie u. Geophysik
                        Universität Frankfurt
                        Feldbergstr. 47
                        6000 Frankfurt/Main 1

Subject Index